The End of the World As We Know It

How and Why We Find Ourselves in the Age of Climate Change

First Edition

To David and Siona,
Congratulations on your marriage. Lots of Love

Oliver Lewis Thompson

For family.
For Sabah.

ISBN: 978-0-9559933-0-5

This book also available from author's own site:
http://www.oliverlewisthompson.com

Contents

Introduction

Imagine it: dry river beds; encroaching deserts; flooded city streets; harder and more frequent tropical storms; a sever lack of fresh water; dry, lifeless soils; abandoned metropolises crumbling with neglect and the passing of time; huge global debt; war; famine; disease and extinction. Is this the world we are heading towards?

Climate change is one of the largest issues on the political agenda today, but how real is it? Is the Earth's climate really changing, and how dramatic will this change be? What are the consequences of climate change and do they really matter? More importantly, is it being caused by us?

For years, all I was hearing in the news and on television was that the point-of-no-return for climate change was 'just around the corner', and that we needed to act swiftly to put a stop to it. Then, all of a sudden, the phrases 'point of no return' and 'just around the corner' became less frequent. Climate scientists began suggesting that the point of no return may have already been and gone, and that all we could do now was soften the inevitable blow as best we could. Though many sceptics dismissed such claims as too 'alarmist' – and many politicians remained muttering in their chairs about the whole ordeal – the gradual passing of time has seen more and more experts agree that our world has already begun changing, whether we like it or not.

When US President Bush came to power in 2000, it suddenly became acceptable again to denounce climate change. Politicians and journalists began questioning the science and started wagging fingers. In forecasting there is never certainty, leaving those with vested interests to point accusingly at tiny cracks in climate change forecasts as if they were gaping holes full of dirty little secrets. "It's all flawed!" they shouted, stamping their feet and thumping their fists. "How can you predict what the world will be like in 2050 if you can't even get tomorrow's weather accurate?" Ingenious to their plan was a huge campaign that convinced the American – and to some extent the European – people that the question of climate change was still debatable, that more than half of all 'experts' questioned the science behind the claims.

In the first few months of gaining power in 2000, the Republican administration cut funding for research into renewable energy resources by 50 per cent, cut research into cleaner and more efficient cars and trucks by 28 per cent, broke a campaign promise to invest $100 million per year in rainforest conservation, cut $500 million from the Environmental Protection Agency's budget, tried to reverse regulation protecting sixty million acres of national forest from logging and road building, pulled out of the 1997 Kyoto Protocol agreement on climate change signed by 178 other countries, and did a whole host of other bad things for the environment.

I noticed all this happening, as did millions across the world, in deeper stages of shock. The trend from a world slowly getting greener to a world questioning whether green was the right colour to be, seemed to be a world gone mad. It was as though we'd all taken a big step into the past and had yet to figure out what exactly these substances were coming out of our chimneys and exhaust pipes.

Then, in 2005, I went for a four-week backpack around the island of Cuba, where I expected to hike, bike ride, swim, scuba dive, and generally explore the beautiful Caribbean island. Unfortunately, upon arrival, my companion and I turned on the television in our little Havana hotel room to find that we were 24 hours away from what the American news channel was calling the biggest July hurricane ever on record – later to be known as Hurricane Dennis. And it was heading straight for Havana.

As we thrilled at the surging winds and crashing waves of the Hurricane from the safety of our evacuation point, thousands of people had had their lives completely torn apart as Dennis tore its way through the middle of the island, devastating Haiti beforehand. We had to cut our adventure short by two weeks because the roads and communications systems between the west of the island and the east had been severed down the middle, in the path of the hurricane, leaving us with no access to the east.

Back at home in the UK - one of the more boring countries on the planet in terms of weather - we watched over the following weeks as the 2005 hurricane season took off in spectacular style, leaving behind a wake of death and destruction everywhere the hurricanes went. It turned out that being in the middle of the biggest July hurricane ever recorded wasn't so impressive, not when hurricanes Katrina, Rita and Wilma entered the game. Katrina famously devoured New Orleans in late August 2005, the largest ever to hit the mainland United States, and Rita, also a category five, followed up, luckily striking less populated areas. Wilma, however, was different; although it struck Mexico, Cuba and Florida and left thousands without clean water and electricity, it is more significant for the fact that it was the strongest hurricane ever measured. The 2005 hurricane season was the worst in all of recorded history. Records were broken time and time again, and nature's anger went on for well into December due to some fearsome tropical storms.

Climate change was back on the agenda, as more and more news agencies decided that there may be something interesting about what climate scientists have to say after all. The Age of Denial continues, but for anyone in their right mind, people who deny climate change are akin to those who deny the Earth orbits the Sun. Slowly, gradually, the voices of the minority will be ignored, no matter how loudly they shout, or how big they make their little club to be. We have already entered the Age of Consequences, and the Earth is already changing... now... today!

When I begun writing this book it was 2003 and still rather unfashionable to talk about climate change. Politicians could certainly talk a good talk but they quickly seemed to forget how to put one foot in front of the other when it came to walking the good walk. The media were hardly interested at all, being mostly caught up with war in Iraq. It started to occur to me that climate change was like a ticking time bomb, and, though we knew how to disarm it, nobody really could be bothered. It just wasn't being taken seriously enough as a global issue and measures and targets set to prevent it were either too low, too distant or being completely ignored. What I really wanted to see was a sense of urgency... and there was none.

And it *is* a global issue. It affects everyone, from farmers in Tanzania to oil tycoons in Texas, from insurance salesmen in Bombay to politicians in Bahrain. Never before in all of human history has a single issue threatened every single person on the planet. Never before in all of the human history have we seen a threat that could wipe our entire species from the face of the planet, along with many others too. Never before in all of the *planet's* history has a single species threatened to disrupt the balance of global nature.

But are the consequences of climate change so bad? I've had that same question posed to me many times whilst writing this book by honest and intelligent people, and I can only tell them the same thing: it could be. In fact climate change could pose no serious threat whatsoever, beyond the chaos it has already caused for low-lying island nations and those afflicted by drought and flooding around the world. A hundred years from now the world could be no different from today, and if it does change it may only be small and manageable.

But it could also be devastating, and we could be looking at the dawning of a new geological Age – and age of the gradual decline of the human from a dominator of the planet to a fringe species, an age that changes the face and the fabric of the entire planet for thousands of years – the Age of Decline, the Age of Extinction… the Age of Climate Change.

Is this alarmist? Maybe the most extreme predictions of a downfall of humanity and the breakdown of Earth's natural ecosystems are unlikely, but if we tie in what science knows of Earth's climatic past to what we are currently doing to the climate, then we see that a millennia-long decline of the natural world is actually quite realistic. Though it is easy to scoff at anybody who tries to predict the unwritten future it is also supremely irresponsible to ignore scenarios that are very real and very possible.

First let me stress that this book is not for academic purposes. When I started writing this book there were few modern mainstream publications that dealt with climate change, and since then, though plenty of new books have come out, I still find it frustrating to lay my hands on a single book that deals with climate change completely, from a wider perspective. Some deal with the future, some with the present; some with deforestation, some with energy production; and some are just filled with pictures and have little explanation. Some skip over all the issues with the odd descriptive narrative of the author's personal experience, but then skip off all too quickly. I wanted to read something that put all of these things together and I was annoyed that so many other people out there could not properly understand climate change without buying four or five books or spend hours and hours roaming the internet for meaningless facts.

I am young, and when I started writing this book I was nineteen years old and had only just begun my university degree. Other climate change books I've come across on my travels are by seasoned geologists, climatologists, journalists and even politicians – people who've been lucky enough to travel to all corners of the planet and see the changing world for themselves, describing it in vivid detail. Unfortunately I haven't had such opportunities, and I am extremely lucky that the one trip I have managed to make for research purposes coincided with the surprise arrival of Hurricane Dennis, the first major tropical storm of the no infamous 2005 hurricane season.

But I am young, and that at least gives me one advantage over those other climate change writers because I have more to lose than they do. I feel more urgency about the issue, more passion, and more frustration than they can ever understand. I feel this urgency not because I have foolishly allowed myself to get carried away with it, but because it is the younger generations who will have to live and cope with the consequences of it. In that respect I am thankful that I am not even younger, and it is for those even younger than me – the infants, teenagers, the unborn masses – whom I dedicate this book to. Those who have to tidy up the mess we make. Those who have to suffer, scrounge, and perhaps die in the new world... because we couldn't be bothered changing the way the world works today.

But this book goes beyond merely skating over the issues. We will look deeper into what climate change means for other species, for human populations and for the future populations. We will look at the growing deserts and the rising seas and see how this has already affected people, let alone how it will affect them in future. We will look at the science behind climate change, looking at the nasty surprises that await us on our bumpy road to a new world.

Finally we will find the root cause of the problem - deeper than the topics such as deforestation and energy production can take us. We will look at the politics of climate change, what has been said in the past and what has been done. Will we finally answer the question: 'what really stands in our way of living in balance and harmony with the natural world?'

This book exists to bring the truth about the greatest issue of all time to the people who matter the most: everybody.

Terminology

Additionally, I have tried, throughout the book, not to use the term 'West', or 'Western' when wanting to describe the cultural, economic or other such values to do with the rich, developed, dominating nations of Europe and North America. This is because the term is old-fashioned, implying inclusion of countries generally in the Western hemisphere, like those of South America which are very different from countries like the US and Europe. The term also suggests that countries like Japan, Russia and Australia are not included, though their economic and environmental undercurrents are incredibly similar to traditional Western countries.

Instead, when talking about 'rich and poor' nations, I have divided the world into two unequal sections. Countries that are seen to be industrially and economically developed often lie well inside the Northern Hemisphere; some more than others have relatively few problems when it comes to everyday quality of life, measured not just by wealth, but also by health and education. The following nations will be referred to as the 'Rich North' indicating their wealth, high standard of living, and their common geographical position: Canada, United States, UK, Spain, Portugal, Eire, Norway, Sweden, Finland, Denmark, France, Belgium, Netherlands, Luxembourg, Germany, Italy, Austria, Switzerland, Greece, Czech Republic, Israel, South Korea, Japan, Australia and New Zealand. Others also falling under this category may not be as wealthy, or may have a larger proportion of the population close to the poverty line, but I have included them under the 'Rich North' name because their quality of life and generation of wealth is still substantially higher than those of the poor. These nations are: Estonia, Latvia, Lithuania, Rep. of Macedonia, Slovakia, Hungary, Slovenia, Croatia, Russian Federation, Turkey, and Argentina (admittedly in the South).

The second division will be called the 'Poor South' – again because of the undeniable fact that many, if not all, exist around the equator or below it. The people in these nations have substantially low quality of life, life expectancy, and access to services such as health and education. Poverty is a widespread plague in many of these countries and they are largely without real world power - more often they are subject to it. The 'Poor South' is as follows: Colombia, Bolivia, Chile, Paraguay, Haiti, Papua New Guinea, Cambodia, Laos, Bangladesh, Nepal, Bhutan, Pakistan, Afghanistan, Yemen, Malaysia, Singapore, Guyana, all of Sub-Saharan Africa, and Palestine.

Many nations cannot be categorised so simply - their overall wealth and quality of life per capita is too high to be part of the Poor South, but they remain not wealthy enough as the vastly different Rich North. Examples include Venezuela, Mexico, Libya, India, Thailand, Indonesia, North Korea, Morocco, and Egypt. Some appear wealthy in terms of the production of capital, but the distribution of their wealth is so uneven that in reality their populations remain impoverished whilst only a few individuals retain the vast riches. Such nations have a significantly lower quality of life than their economic wealth suggests and stand out for it: Saudi Arabia, Oman, Qatar, Algeria, Tunisia, Dominican Republic, and Iran. On the other hand, other nations in this 'none-category' have substantially higher standards of living than their economic wealth suggests: Mongolia, Philippines, Burma, Lebanon, Georgia, Ukraine, Suriname, and Cuba. These 'none-category' nations sometimes fall into the Rich side, and sometimes the Poor side, depending on what demographic is being discussed; throughout this book these nations will usually be referred to individually if they significantly stand out.

Finally, throughout I have used the British version of the value of 'billion'. Generally the consensus is that 1 billion is equal to 1000 million (1000, 000, 000) but there is a second school of thought that likes to use 1 million million (1000, 000, 000, 000) instead and this is what I find more appropriate – simply a personal preference. Therefore, I am able to state numbers usually referred to as billions as a more accurate 'thousand millions' and hence avoid any confusion amongst people who do not know which to expect.

PART 1 - Planet.

We live in an amazing place. On the face of it, Planet Earth is nothing but a huge ball of molten rock spinning around a giant nuclear reactor. At the centre of the Earth is a half-liquid, half-solid mass of exceedingly hot rock known as the Core, surrounded on all sides by a viscous liquid Mantle, and covered in a very thin layer of solid rock Crust. Above the surface of this crust is an even thinner layer, comprised of a wild concoction of gases, dissipating off into the nothingness of space some 140km (90 miles) above – what we call the Atmosphere.

Earth is one of eight planets orbiting its nearest star, the Sun, (the ninth, Pluto, was demoted to the new classification of 'planetoid' in mid-2006 by an international symposium of cosmological scientists) and is the third closest at about 149.6 million kilometres away. Earth remains the only planet with an atmosphere calm and dynamic enough to sustain complex organic life – it could have easily ended up like Venus, a boiling cooking pot of carbon dioxide that averages 480°C on the surface at night, or like Mars, a wispy world ravaged by planet-wide dust storms and unprotected from intense heat in the day and freezing cold at night.

But those are only the basic mechanics of it all. What is *really* amazing about Earth is that it is probably the *only* place where you could find life forms in this solar system – possibly even the entire universe. Not only that, but life on Earth is also incredibly complex, unfathomably diverse, and it seems to flourish wherever it gets even the slightest opportunity. Teams of scientists would find it impossible to sustain bacteria out in the open on Mars, but leave an apple core lying around on Earth and in just minutes colonies of bacteria will be turning it brown.

It is the favourable atmospheric temperature range described above that allows H_2O to exist on the planet in all three of its physical states: solid (ice), liquid (water) and gas (water vapour). These three substances play a central role in maintaining not only complex and evolutionary life on the planet, but also the complex and evolutionary *climate* of the planet. And as soon as we start looking deeper into life on Earth, and the dynamic climate system, it becomes clear how much more there really is to this ball of molten rock.

The existence of life on Earth is astounding. The origin of life probably dates back to around 4000Ma (million years ago), and primitive life forms like blue-green algae have been found in rocks as old as 3900Ma. This is quite surprising, especially when you consider that the Earth is no more than 4500 million years old, and for the first 500 million years of its existence it was being constantly bombarded by debris from space.

Despite this flying start, life on Earth only got moving at a very gradual pace, and for the first half of its history it barely evolved beyond organisms more complex than small lumps of cells. In fact, it is around 1000Ma when we see the first traces of multicellular life forms appear. By 700Ma primitive animals such as jellyfish had appeared. The second half of life's existence has since proven to be full of interesting occurrences, developments and mass extinctions that make the first half look incredibly blank and boring. We have the appearance of the first fish, the first insects, the first birds, the first flowers and so on.

In fact, scientists use the changes in the history of life as milestones for looking at the past. They do this by highlighting several things: mass extinctions, local extinctions, population explosions, and development of certain species types. Using such milestones they can break down the history of the planet into smaller time periods – making study more digestible.

The longest time periods are called Eras and there have been six of them altogether. The one our species currently dwells in is the Cainozoic, which is currently 65 million years old. Eras mark times when massive changes have occurred on the planet and where as much as 95 per cent of all species have become extinct over short periods. They signal the biggest, baddest moments in the planet's chequered history - the last Era change being the moment a massive asteroid collided with the planet, signalling the end of the Age of Reptiles and deleting the dinosaurs and many other creatures. But (as a matter of interest) it wasn't the actual slam of the asteroid colliding with Earth that caused this great mass extinction - the deciding factor was the climate. The collision sent a large volume of material into the atmosphere, blanking out the sunlight and acidifying water supplies. This triggered a climate side-step, plunging all unsuspecting organisms into extinction and leaving only the hardiest alive to carry the torch.

The Era of dinosaurs was actually called the Mesozoic but is usually cut into smaller chunks called the Triassic, the Jurassic, and the Cretaceous.

These smaller pieces of history are called Periods. We currently reside happily in the Quaternary period - around 1.8 million years old and marked, not with any modesty, at the estimated appearance of the first humans.

Even smaller than periods are Epochs, which pinpoint times of large changes in species dynamics, such as when there is a large influx of new species into an area, and a sudden disappearance of the species already there. These migrations can be caused by the appearance of land bridges (caused by low sea level), continents coming together (caused by plate tectonics), or large climate changes that heat or cool the planet and affect how far species can migrate. The human being currently wallows in the Holocene epoch, which has proven to be a period of calm, modest climate, and has seen humanity advance from foragers to farmers, farmers to factory workers, and factory workers to fossil fuel addicts.

Today Planet Earth is a living breathing organism - complex, intriguing, still not fully known and always wonderful. Everywhere life is teeming: from the depths of the lush, green jungles to the cold permafrost of the Arctic tundra - from the tips of mountains to the black depths of the deepest oceans. There are fish as big as trucks, birds that can fly faster than your average train, creatures with eyes as large as dinner plates and trees taller than a tower block, with trunks wide enough to drive your car through. We currently know of over a million different species of organisms on Earth, from the biggest (the Giant Red Wood or *Sequoia*) to the smallest single-celled organisms that are literally everywhere, and there may be as many as ten million yet to be discovered.

Chapter 1 – Climate: the Fundamentals of a Dynamic System

Apart from being the only planet in the solar system with extensive life, Earth is also the only one with a complex, systematic climate capable of supporting that life. This is down to a number of reasons; the most influential being that Earth is the only planet orbiting the Sun within an optimum *distance* for life to exist. Venus has a dense atmosphere, as does Titan, one of Saturn's moons. Jupiter's moon, Europa, probably has liquid water beneath its icy surface, and the largest of the Jovian satellites, Io, has active volcanoes even at present. There are even traces of river beds all across Mars. But while these other solar bodies can claim to have Earth-like climatic qualities, only Earth has all these qualities *and more*, blended spectacularly together to make a dazzling array of dynamism and wonder; an oasis in the infinite nothingness of space.

Components of the System

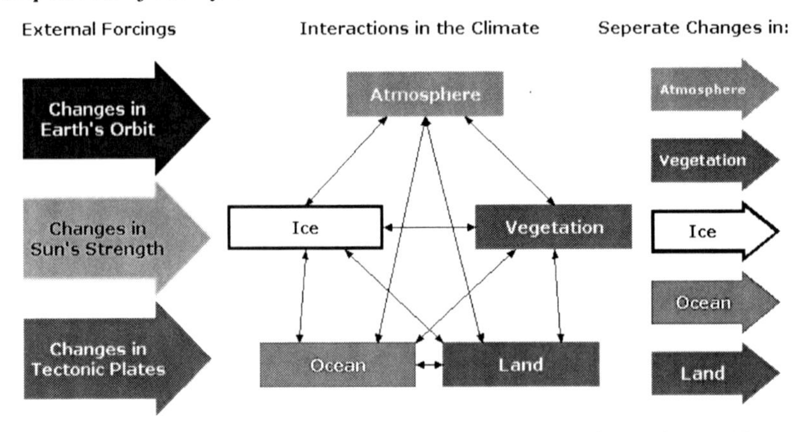

Figure 1.1 – Only a small number of external factors force natural climate changes. They cause changes in the planet's internal components, forcing them to push and pull each other to find a new equilibrium. As a result, each of the components changes.

Earth's climatic system consists of air, land, ice, water and vegetation. Changes in these components act in a basic cause-effect relationship – something causes change and something else is affected. These processes work all over the planet, from the frozen poles to the sweltering tropics.

Without this complex combination of key components and vital processes, the Earth's climate would be unrecognisable, and would be without the ability to create and support even simple life-forms. As you can see from the middle of Figure 1.1, every component of the climatic system interacts with every other component.

Air is more than just the blue-white stuff we see above our heads every day. We can feel its warmth without having anything to hold on to or touch. Our nerves sense it all around us, all of the time, yet we cannot see it or grasp it in our hands. It is a liquid without being liquid. It is enormous, but invisible. It is clear one day, full of clouds of water vapour the next. And it is so complex that it can be divided up into four distinct layers, each defined by its temperature and the direction of its temperature gradient. Humans can breathe successfully in the lowest five kilometres of the first layer – the Troposphere – which actually extends to around 12 kilometres above sea level. For an idea of scale, consider that Mount Everest rises only 8.8km above sea level.

The troposphere is special because it is the only layer firmly cut into two around the equator, whereby air currents in the Southern Hemisphere hardly ever mix with those of the Northern Hemisphere, and vice versa. Secondly, it is here where the warmest air tends to rest close to the planet's surface rather than higher up, seemingly contradicting common sense (heat rises doesn't it?).

The other three layers are the Stratosphere (home of the mighty Ozone layer, which guards against ultraviolet radiation from the Sun, effectively converting it to heat), the Mesosphere (-90°C in temperature), followed by the Thermosphere, where gas molecules gradually become less and less abundant until they finally peter off into the blackness of space. Here temperatures may reach 1000°C, but there are so few air particles that you couldn't actually *feel* this heat. Together the troposphere, the stratosphere, the mesosphere and the thermosphere make up the atmosphere – the thinnest, most volatile layer of the planet and the one we owe our complete existence to.

Surprisingly, the most abundant gas on Earth is nitrogen. This is surprising because we live on a planet flush with life forms, blue skies and high misty mountains and we usually associate these things with oxygen. Nitrogen comprises around 78 per cent of the atmosphere, with oxygen coming in second at 21 per cent. The third most abundant is even more surprising: argon

– notching up a total of 0.9 per cent. All room for error considered, we could say that these three gases make up 99.95 per cent of the air we breathe.[i]

The remaining gases are many, and seem to set the fashion for the time. They may only exist in tiny fractions (even collectively) but they manage to have a huge force in world climate, leaving the big three with a lot to answer for. Ozone is a famous 'trace' gas, consisting of only ten molecules in every million floating around up there, yet its effect is large enough that it acts as a 'second skin' to life forms on the surface, protecting and preventing harmful solar radiation from causing illness and death. In concentrations greater than just ten molecules per million, ozone at ground level can be deadly for animals to inhale. Indeed, ozone at the surface is one of the growing problems facing people in smog-filled cities around the world.

The most powerful gases of the atmosphere are the Greenhouse Gases, just more than 30 in total, of which carbon dioxide (CO_2) is the most famous and important (the concentration of which we will find out later because it is central to the entire book). Methane is the second most important but, although it is sixty times more effective at trapping heat than CO_2, it isn't as abundant, comprising just 1.5 parts per million (ppm) of the atmosphere.

Unfortunately, this concentration has doubled over the last few centuries, no doubt the result of a growing human population here on Earth; methane is made by tiny microbes that live in bowels of mammals and stagnant pools of water, so more people means more breaking wind, which is the principle way humans like to emit methane. Thankfully, methane doesn't hang around very long in the atmosphere, especially compared with CO_2. Still, it is known that the gas has exacerbated many a climate shift in the past through 'feedback' processes - it actually makes bad situations worse through the way it interacts with the rest of the system.

Nitrous oxide is 270 times a more effective greenhouse gas than CO_2. Created by burning fossil fuels, bushes, trees and grasses, and using nitrogen-based fertilisers, nitrous oxide is an ever-increasing threat. Also known as 'laughing gas', nitrous oxide has increased in concentration in the atmosphere by 20 per cent in the last couple of centuries - making it not such a laughing matter after all.

Essentially, there are three fundamental kinds of climate forces in the natural world[ii]:

1. **Tectonic processes generated by the planet's internal heat**. The processes of 'plate tectonics' are very slow and very gradual but are constantly in a state of change.
2. **Changes in the planet's orbit.** The Earth does not orbit the Sun in a perfect and unchanging circle but does so in a varying ellipse shape. It also changes its angle of tilt, as well as the tilt direction.
3. **Changes in the strength of the Sun's energy output.** This also affects the amount of solar radiation received by Earth. The major fact to remember is that the strength of the Sun has been increasing throughout Earth's 4,550 million-year lifetime. In addition, there are shorter-term fluctuations that occur over decades, centuries and millennia that are partly responsible for global climatic changes.

Response Time for Climatic Change

There is something we must appreciate before we can begin to learn about the climate as a whole: each process of the system is never constant and all vary in strength, direction, volume, etc. For example, snowfall is never just snowfall - it can be very light or very heavy; the wind changes its direction and its strength all the time. We can never accurately predict any of these systems because they are almost sporadic and appear quite random, though they are not.

It is important to bear this in mind because one of the first lessons to learn about the climatic 'force/response' relationship is that there is a delay process involved. A useful way to imagine this is to picture a flame being lit below a kettle of cold water – although the flame is incredibly hot, the temperature of the water doesn't suddenly jump to this temperature in an instant; it takes a certain length of time for the water to gain energy and boil. In a similar way, there is a delay period between clearing a forest and then seeing the knock-on effects of this in the atmosphere and the land surface.

Each of the major components of the climatic system has its own characteristic response time and these can range from hours and days up to tens of thousands of years.[iii] The fastest component to respond to temperature change is the atmosphere, as anyone will know from waking up to a warm early morning, even though the Sun has only been above the horizon for a couple of hours. The surface of the land is comparably slower but still shows response times that are within hours, days and weeks. Beyond this is water, in its many forms over the Earth: the wind-stirred ocean surface, and smaller bodies of water such as lakes, may react to temperature changes in only a few days or weeks. Yet the sluggish deep ocean lies far from the manic atmosphere and may respond only after hundreds or thousands of years.

The slowest of the components is the ice, and although the surface ice shelves on a polar ocean can melt or grow within periods of months the major changes in ice sheets and glaciers take decades or centuries. The Antarctic ice sheets usually only show full change after a minimum of ten thousand years, depending on the exact degree of temperature rise. Vegetation differs somewhat from the other factors already discussed. An unseasonable frost can kill off leaves and shoots overnight, while pioneering forests, occupying newly-exposed ground can take thousands of years to reach a full potential. In other words, if temperatures suddenly shot up tomorrow you'd feel it in the air and see it in your potted plants, but unless those temperatures stayed that way for a long, long time, the deep oceans and the bigger ice masses wouldn't really care. Hence, when temperatures change, the full effects may not be seen straight away.

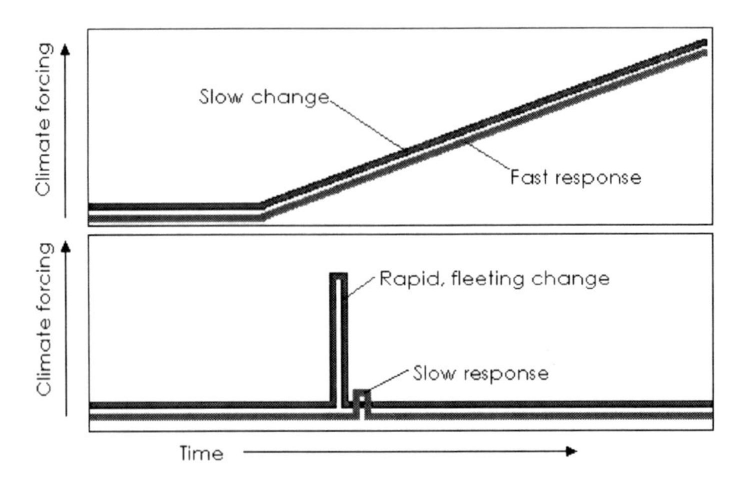

Figures 1.2a and 1.2b (above) – There is a delay period between forcing and response. Gradual changes often allow for fast responses and reduce this delay period (a); but if the forcing is too fast the delay period is increased, and if the forcing only occurs for a brief period there is little chance for a full response (b). (Adapted from Ruddiman, W. F. (2001). *Earth's Climate: Past and Future*. W. H. Freeman & Co., New York. Pg 13.)

The first graph above (Figure 1.2a) is an example of the how response time changes at the same rate to the acting force and shows hardly any delay period at all. The change is occurring almost immediately because the force is acting very, very slowly. If the force were to act any quicker the delay period would increase and the response would not happen at the same rate. The extreme of this is shown in Figure 1.2b where the force acts very quickly but only very briefly - the result is a minimal response, inducing hardly any change at all. To picture these at work, imagine the central heating system of a house. In Figure 1.2a the thermostat is being turned up very slowly over a long time period, inducing the house to warm accordingly. In Figure 1.2b the thermostat is turned up high but only for a few minutes – the temperature of the house hardly increases.

The existence of delays in the climate system provides it with a safety net against wild, short-lived swings. Gradual changes are followed closely but fleeting changes are not. On the other hand, this cushioning effect can also hide the true wrath of the forcing - unleashing the response at a later date when the forcing has already disappeared. Bear all this in mind later, when we begin to look into present-day climate forcings… and future climate changes.

---

[i] *Gaseous makeup of atmosphere*: Flannery, T. (2006). *The Weather Makers*. Penguin Books Ltd, London. Pg 21-22.

[ii] Ruddiman, W. F. (2001). *Earth's Climate: Past and Future*. W. H. Freeman & Co., New York. Pg 10.

Chapter 2 – A Brief History of Ancient Climate

Earth's climate is always changing. A hundred million years ago Earth had no Polar ice caps, yet at other times throughout its history the ice caps have been so large that they engulf much of the northern hemisphere and large parts of the southern. Also, we know that the continents have been moving around the surface of the planet for at least the last 200 million years (if not since the moment the planet's molten surface cooled solid in its earliest years) - as a result there have been times when the entire land surface has been covered in rich vegetation, and when yesterday's Antarctic was home to life as diverse as we find in today's tropics.

All of these changing forces have pushed and pulled at the planet's climate since it was first conceived. If we really want to understand Earth, if we really want to predict the future climate and assess the potential impact of global warming, then we must delve deeply into the planet's dark and misty past and see exactly how things were, and why they were that way.

When it comes to climate there is an awful lot of talk about greenhouse gases, and one in particular: carbon dioxide. But just how important is carbon dioxide to the climate? Why does it increase and decrease its volume in the atmosphere, and what causes it to do so?

Presenting the World Famous CO_2

As we already know, life can flourish almost anywhere on this planet, aided greatly by its liquid water and its favourable mean temperatures. But why has Earth been able to support life for most of its 4,500 million year history?

Earth's nearest planetary neighbour, Venus, has an overall chemical composition which is very similar. It orbits only 72 per cent as far from the Sun as the Earth, but its mean surface temperature is a whopping 460°C. Its atmosphere is made of a thick layer of sulphuric acid that reflects 80 per cent of its incoming solar radiation – much more than Earth, which reflects a quarter. Furthermore, Venus reflects 515 Watts (W) per metre squared (m^2) of its total 645 W/m^2, absorbing only 130 W/m^2. Earth on the other hand receives 342 W/m^2 of radiation but reflects just 100 W/m^2 - therefore allowing 242 W/m^2 to be absorbed.[i]

So if Venus sends most of its solar energy back into space, how come it is so much hotter than our home planet? The answer lies with Venus' thick atmosphere, which is around 90 times as dense as our own; around ninety six per cent of that atmosphere is carbon dioxide.

The radiation reaching Venus' surface becomes trapped under this thick blanket of CO_2 and is retained as heat. Earth's ability to trap radiation is comparatively poor because its composition of atmospheric CO_2 is a tiny 0.02 per cent. All this goes to show that Earth's small greenhouse gas effect *combined* with the planet's fortunate distance from the Sun, creates a world lush with organic life. Venus, on the other hand, is the hottest planet in the solar system.

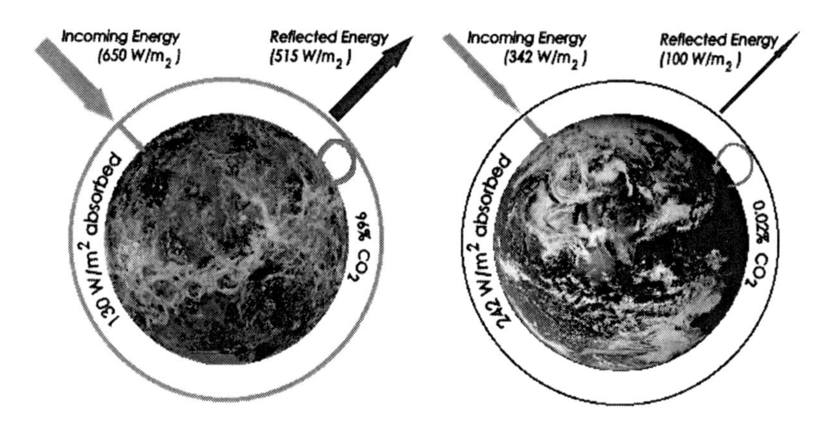

Figure 2.1 – Though Venus (left) receives almost twice as much solar radiation as Earth (right) less can penetrate the atmosphere thanks to its dense cloud cover. However, Venus is still hotter than Earth because it has higher atmospheric CO_2 levels and therefore a stronger greenhouse effect.

One of the major reasons that the global temperature is so pleasing to life on Earth is because our planet happens to be a fantastic distance from our nearest star – the Sun. But the Earth was born when the Sun was in its infancy, and as the Sun has matured, its strength has increased by 25-30 per cent over Earth's lifetime. This is a massive change in energy output, yet the planet's oceans have never boiled away, turning us into a dusty rock like Mercury or Mars – why? This is a climate paradox.

There must be some natural 'thermostat' enabling the planet to remain within the boundaries of habitability without going too cold or too hot – like the thermostat in your house. Climatologists know that if the Sun were any weaker, even by just a couple of percent on today's level, all the water on the planet would freeze – lakes, oceans and all. The sheer weakness of the Sun's strength would override the natural greenhouse effect and cause a planet-wide freezing, damaging all life on Earth. Furthermore, climate models have shown that with such a weak Sun and with greenhouse gas levels at their present value, the early Earth would have remained frozen for the first 3,000 million years of history. Unfortunately, this theory completely disagrees with the Earth's own climate record which shows that the planet has *never* frozen over completely. So what's going on?

Thus we have our paradox. This mystery is known as 'the faint young Sun paradox.' The climate records show that the interval between 850Ma and 550Ma saw the Earth cool down so much that it became what is known as 'Snowball Earth' – almost completely frozen over. But this does not explain why the Earth was not frozen for the several billion years *preceding* this point when the Sun was even weaker still.

The answer lies with those greenhouse gases and their special property of trapping incoming radiation and keeping Earth warm. Modern-day concentrations of greenhouse gases are too low to counteract the low energy provided by the young, weaker Sun, but these gases were much more abundant during the early stages of the planet, providing a thicker blanket to keep the planet warm. To offset the low level of solar luminosity early in Earth's history, it is estimated that there would need to be 300-3000 times more atmospheric CO_2 than at present, and since scientists know that the carbon locked in Earth's rocks today was once in its atmosphere, and easily could have provided such concentrations, so we have the answer to our paradox.

Carbon dioxide is the most common of the greenhouse gases and therefore the one with the most authority. Having more of it in the atmosphere during the planet's early years kept the planet warm while the Sun was still weak. As the sun has grown stronger over time the amount of carbon dioxide has compensated for this by gradually falling. Now all we need to know is what, if anything, told it to fall.

So what's the Thermostat?

The carbon stores on Earth today exist in every component of the system: the biosphere, the atmosphere, the hydrosphere and the lithosphere. Most exist in rocks and sediment beds of the lithosphere and relatively little exists in the atmosphere. More importantly, the size of the carbon store indicates how quickly it releases its carbon to the cycle – in other words, transferring the carbon from one store to another. The smaller stores of the atmosphere and the biosphere are quick to release their carbon, whilst the largest stores like rocks take a vast amount of time. If you imagine this transferring process as a cycle you can see why the carbon 'builds up' in the places where it takes longest to pass through, rather like traffic-jams in the inner city and empty roads away from the city.

The remarkable thing about carbon is that it is always on the move – it never stops cycling between the different components and transferring to different forms. Carbon is the fuel of the planet - the one element that can be recycled and used over and over by the Earth for doing almost anything; from building rocks to building life, carbon is there. Carbon can crop up anywhere really, depending on what other elements or molecules it has bonded with. Even carbon in its purest form can be unrecognisable, from the blackest soot to the shiniest diamond - its form simply depending on how its atoms are arranged. Philip Ball points out in his book that there is so much water on Earth that we ought not to name our planet 'Earth' but rather 'Water'.[ii] In a similar alternative, perhaps a better name would simply be 'Carbon', just out of sheer respect.

Naturally the carbon in rocks has two gateways for passage into the atmosphere: volcanic eruption and oxidation of organic carbon in sedimentary rocks. Far beneath the surface of the Earth, beneath the seat you are sitting in right now, is a tremendous and immense sea of boiling, hot, liquid rock. As the hot iron core radiates heat upwards it warms the liquid mantel, and this in turn boils the crustal rock above it. It takes an incredibly high temperature to turn rock into mush, but the interior of the Earth is between 4000 and 7000°C and this is just about right.

It doesn't take a genius to work out why volcanoes exist. If you boil water in a pan and leave the lid on, you will start to see the lid moving and being forced upwards as the hot steam tries to escape. If you tied this lid down so that it cannot move and cannot let the heat out, the result would not be very pleasant, or safe to witness. The Earth has the same principle - all the gathering heat below the rocky crustal surface would build and build into something quite dangerous if it did not have a means to release some of this energy and ease the tension. Volcanoes are the result of this tension building up, and at any weak spot on a surface they will eventually puncture and gracefully explode.

For the first half of its history, Earth was highly volcanically active, creating a large pool of carbon in the atmosphere. As the planet cooled the number of volcanic eruptions eased up, becoming the rare events we know today. So if there were more volcanoes during the planet's early years than there are today – implying a slow decline – then surely this had gradually eased the amount of carbon in the atmosphere and hence been the thermostat compensating for the growing strength of the Sun. As one went up the other went down. Is this the case?

Although it may seem likely, the answer is no. The processes of the interior of the planet operate without influence from the climate system and climatic temperature changes affect only the uppermost layers of the Earth's crust (to about ten or twenty metres below the surface). By definition, a thermostat mechanism has the ability to respond to changes and adjust itself. Volcanic output does not have the ability to respond because the bubbling magma miles below your seat is not sensitive to changes in the air... and if it were, it probably wouldn't care.

Of course, if all this carbon is moving from the Earth's interior into the atmosphere then it must also be *removed* from the atmosphere to prevent it building up. Over land, this is done by a process known as chemical weathering. In simple terms the carbon dioxide is absorbed and incorporated into soil groundwater, where it forms carbonic acid and begins to eat away at the rock; the tiny pieces of rock then wash away with the groundwater and eventually into the ocean. This is a very slow process but is undeniably persistent; recent studies have shown that it accounts for 75-80 per cent of the

0.15 gigatons of carbon buried in the oceans and the land every year. Chemical weathering is essentially rocks getting worn away.

So is the weathering of rocks the natural 'thermostat' of the planet? We know that natural chemical weathering occurs more rapidly in higher temperatures than in lower ones. Controlled experiments have shown that weathering rates double with every 10°C increase in temperature. Precipitation also influences weathering and rainwater is a vital part of hydrolysis. The relationship between latitude and precipitation is similar to the relationship between latitude and temperature. Just as the tropics are the hottest of the zones on Earth, they are also the wettest – just like the Polar Regions are the coolest places on Earth, they are also the driest in terms of rainfall. Temperature and precipitation are therefore very closely linked. In fact, the capacity of the air in holding water vapour rises with its temperature, again doubling every 10°C increase.

However, some areas are closer to the tropics than others, yet they experience far less rainfall. This is because the relationship between latitude and precipitation is not perfect, and why it rains more in Europe than it does in North Africa. This is down to the third and final environmental factor that controls chemical weathering: vegetation. All plants take CO_2 out of the atmosphere and use it in their extraordinary life-giving process of photosynthesis. Photosynthesis is a truly marvellous process that non-plant organisms can only envy.

Plants absorb sunlight, water and draw in carbon dioxide from the surrounding air. Photosynthesis uses the carbon to create new cells for the plant and disposes the remaining oxygen back into the atmosphere. When plants die, the carbon in their cells is delivered to the soil, combines with groundwater and forms carbonic acid, or H_2CO_3. Though this acid is weak, it still enhances the rate of chemical breakdown in rocks and minerals.

Vegetation is closely linked to precipitation and temperature. Moist, tropical rainforests are thick and lush with vegetation, whereas the drier savannahs are much more sparsely populated plant species. Additionally, the closer to the equator you get the hotter it gets, and hence more solar energy is available for plants to photosynthesise.

In 1981, James Walker, Paul Hays and James Kastings proposed that there is a climate-dependent 'negative feedback' due to chemical weathering. This meant that the Earth's temperature was kept in check by the chemical weathering process, a process triggered or halted by climate extremes. We have to create two extreme hypothetical situations to test whether this process is the thermostat of the Earth, and see how this negative feedback works.

To begin with, we can imagine a situation whereby the Earth's climate gradually warms into a 'greenhouse' state. A warmer climate only encourages more vegetation growth, as well as more moisture in the air. Increases in both begins to speed up the rate of chemical weathering of the land, and hence the rate of CO_2 removal from the air. The result is a reduction of the initial warming. In the end it is the very increase in temperature that leads to its own decrease. Similarly, any cooling of global temperatures results in fewer and less dense vegetation hotspots, and far less precipitation. The end result is a net reduction in the amount of CO_2 removed from the atmosphere, and thus a gradual slowing of the initial cooling process.

But, this opposition to the warming and cooling processes does not prevent them from happening in the first place. Because of the delay in the climate system, the changes are allowed to continue for some time before the negative feedback processes initiate. In other words, the chemical weathering process cannot keep the Earth's climate ticking along at a constant and steady rate, but it can prevent it from spiralling out of control. This is, in effect, a thermostat.

Weathering relies heavily on the availability of exposed, surface rock. If we start with a large cube of cheese one cubic metre in size then the total surface area of the cube is six square metres (six sides all one square metre each). Slicing the cube into equal halves along its three major axes creates eight smaller cubes, each half a metre wide, tall and deep. The total surface area of each of these cubes is $12m^2$. The cheese now has twice as much surface area as before but has not lost any of its original volume. If we continue to cut each little cube in half each way the amount of surface area exposed to the air shoots up, but the mass never changes.

After only ten sequential halvings of the cheese over 1 thousand million cubes (each 1mm on a side) are produced, providing a total surface area 1000 times bigger than the original block. Even now, however, the cheese retains its original volume. If we replace the cheese with rock we can see how chemical weathering would increase by a factor of 1000, just because of this single effect. Huge increases like this from such little fragmentation could far exceed the overall effect of temperature, precipitation and vegetation.

Making the climate constant and eternally stable is rather like trying to stand an egg on its most pointed end along a raised plank of wood; you can prevent it from tumbling too far left, or too far right – and shattering all over the floor – but try as you might you will never be able to stand it perfectly still. The thermostat of the Earth works in the same way to the person trying to balance the egg on its thinnest end – it prevents complete breakdown but can never achieve equilibrium for more than a brief moment.

Mother Gaia

Though chemical weathering affects the climate substantially (responding to climate swings) it isn't a process entirely independent in its own right. In fact, a rather significant proportion of the carbon that moves through the weathering cycle does so thanks to organic interference. There is even a sub-cycle for organic carbon acting within the larger carbon cycle.

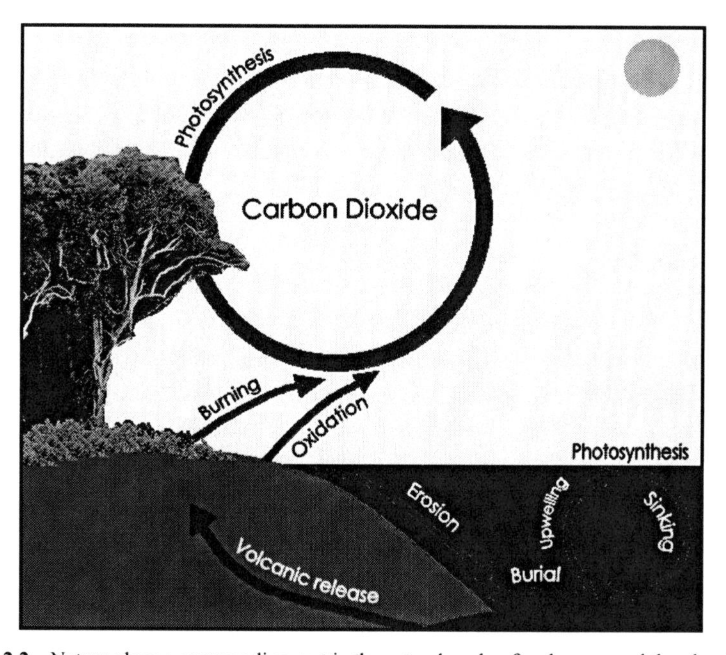

Figure 2.2 – Nature plays a commanding part in the natural cycle of carbon around the planet.

A surprising amount of the total carbon on Earth exists in organic form at any given time – more than a fifth to be exact – and a large amount of organic carbon isn't present in the forests and beasts that you might first imagine; one of the largest organic carbon stores is the ocean. Another is in a place nobody would initially imagine: leaf litter and plant roots are consumed by literally billions of micro-organisms in soil, and subsequently the average forest floor may contain as much carbon as the trees above it.

The biosphere has such a strong influence over the carbon cycle of the planet because nearly *all* carbon stores on Earth are related to living organisms, either directly (soil, coal, oil, natural gas, limestone, forests, oceans) or indirectly (atmosphere, deeper layers of the lithosphere). Photosynthesis is the engine of the organic carbon sub-cycle because without it there could be no movement of carbon through vegetation and some photosynthesising sea creatures, and thus all life.

Since the sub-cycle carries one fifth of the total carbon on Earth, it has a powerful say over the balance of CO_2 in the atmosphere. Furthermore, the organic carbon sub-cycle has the potential to act much quicker than the larger inorganic carbon cycle. Under certain conditions, large amounts of carbon can be extracted from the atmosphere over a very short period of time and this has the potential to alter the climate system more abruptly than the slow-acting chemical weathering process ever could.

However, as an interesting tangent of thought, it should be understood that there are more elementary cycles going on all the time, in all corners of the Earth. All of the essential elements of the Earth – nitrogen, carbon, oxygen, Hydrogen, Sulphur and Phosphate – have a closed cycle on the planet. Each cycle has a biotic phase (organic) and an abiotic phase (inorganic or 'geochemical'). In other words, each of the above elements finds its way in and out of some form of organic life, whether it is plant, animal, fungi etc. As a consequence, organisms have a larger influence over the basic elements of our planet than one might first consider.

Humans have developed a knack for disrupting these natural cycles, particularly the Carbon and Nitrogen cycles, which have such a massive influence over local and global climate when disrupted on a large scale. Furthermore, when considering the previous example of soil, new theories have arisen that contradict common-sense approaches to removal of atmospheric carbon dioxide. You'd think that planting trees would help sequester atmospheric CO_2 and thus help cool the planet. Initially this would happen; however, some scientists think that under higher temperatures, soils will release extra carbon from leaf litter and roots back into the atmosphere, and stimulate global warming. Also, whenever we clear or burn forests today (under cooler temperatures), not only are we releasing CO_2 back into the atmosphere from the trees but we are also removing the need for carbon to exist in soil – effectively cutting down *two* carbon stores at once. Felling trees therefore causes warming and, if we try to plant new trees to take back the carbon, the higher temperatures will only force more to be emitted. It seems counter-intuitive to say that we shouldn't replace trees that are felled, but perhaps this only shows us that felling trees in the *first* place isn't the best idea under the circumstances of a warming world. This point will come back to haunt us later in the book.

Life adjusts to seasonal change. For example, mammals like hedgehogs and badgers hibernate in the coldest months of the year and come out of hibernation in the spring. Schools of fish and pods of dolphins are often seen travelling hundreds of miles through the water to reach more comfortable temperatures during the hotter and colder months, and similar migration patterns occur with birds and butterflies. Changes can occur physically, too: trees in more temperate zones shed leaves in the autumn to prepare for the inevitable freezing temperatures to come; animals with fur have much fluffier coats in winter to hold in more body heat, and then in the warmer months the fur clings to the body to keep the animal cool.

Maybe, some think, the survival instinct of life that drives adaptation may also exert influence over the climate to keep it habitable. In 1979, James Lovelock solidified this theory in his book 'Gaia', suggesting that Earth was one huge organic life form that naturally had self-interest to keep the climate moderate. At first Lovelock's theory suggests some sort of conscious-cooperation is going on between different life forms in a global plot to control the weather. In reality, the theory is trying to emphasize the individual roles each life form plays in maintaining its own existence through its environment, including the atmosphere. Plants are obviously the biggest players in this role.

As was mentioned earlier, one of the biggest organic carbon stores is in the oceans, and of all marine organisms none has a greater influence over CO_2 than plankton[1]. There are many types of plankton in the oceans, some photosynthesize and some don't; however, all use carbon.

Phytoplankton is the photosynthesizer, and is found throughout the oceans, seas and lakes of Earth - their abundance depending on the availability of light and nutrients such as iron. Like all photosynthesizing organisms, phytoplanktons require carbon, which they take from the upper surface waters of the ocean. Foraminifera (forams) are single-celled, shell-covered plankton between a millimetre and a centimetre in diameter and an example of a type of plankton that don't photosynthesize. They add new chambers to their shells as they grow, consuming more carbon. At death, like most marine organisms, they sink to the bottom of the water, which in a deep ocean can mean several miles from the surface. It is thought that around a third of the ocean floor is formed from the ooze of foram's shells, eventually creating limestone several millions of years down the line.

Charles Darwin, the father of the evolutionary theory, postulated that life evolved in order to increase its own reproductive chances of survival. Lovelock's Gaia[2] Hypothesis goes beyond Darwin's evolutionary notion because it suggests that all evolution has happened to have a greater good of the whole planet, such that it produces a succession of life forms needed to keep the planet habitable. In other words, because the early Sun was much weaker than today, early life forms didn't need to take much CO_2 from the atmosphere and were therefore rather simple. As the Sun gained in strength, more complex organisms were needed to take more CO_2 from the atmosphere. This is perhaps why organisms have generally become larger and larger throughout time (between the major extinction events) because larger living things require greater amounts of carbon for building cells and creating energy. Effectively, the maturing of the sun has driven – partially at least – evolution of life on Earth, via the planet's climate.

---

[1] Name 'plankton' derived from the Greek word 'planktos' meaning wanderer or drifter. This is quite an apt name since plankton only move where the ocean tides take it.

[2] Gaia (variant of Gaea); *Greek Mythology*: (n.) – the goddess of earth, who bore and married Uranus, and became mother of the Titans and the Cyclops.

Critics point out that the role the biosphere plays in the Earth's climate today is a fairly recent development, and that early life-forms probably had little or no influence over the overall carbon cycle. How did the very first organisms influence a whole planet? Climate regulation must have been achieved through mostly volcanic and weathering, rather than biological means. After all, for ninety per cent of Earth's history the type of life on the land was too primitive to play a central role in the climate system.

Photosynthesis probably didn't begin until 2,300 Ma and life remained very simple and very basic for about two thousand million years. During this interval, life probably had very little influence over the climate but as time passed, and more organisms gradually gathered in the sea, their influence over the climate grew. As the planet's volcanoes slowly calmed down, this influence turned into dominance. These conditions encouraged the development of more complex organisms onto and across the early landmasses. Volcanoes may have brought Earth back from the brink of Snowball Earth 600Ma, but since then organic life has really taken over the driving seat. This may even suggest that life on Earth will one day have such an influence over the climate system that the climate remains almost completely constant (like successfully standing the egg on its pointiest end!)

Thus is the story of Earth in its first four thousand million years completed. As the molten planet cooled, it formed a dense atmosphere of CO_2, which it held on to thanks to gravity. This carbon came from within the Earth itself through extensive surface eruptions and energetic tectonic activity. Then life developed and spread into the newly formed oceans, developing a greater influence over the atmosphere and the chemical-weathering processes controlling the climate. As the sun's heat output increased over the next few thousand million years, life began to cool the atmosphere by increasing the rate of chemical-weathering and learning how to photosynthesize.

The ultimate thermostat of the Earth's climate looks to be life itself. Why else would the planet need a thermostat mechanism to remain within moderate and *habitable* temperatures? Would the rocks or the water molecules, or the numerous different elements in the atmosphere, really care if the Earth were frozen over or scorched? Surely the only party that has an interest in the climate, let alone the ability to respond to climate changes quickly, is the living, breathing biosphere.

But what, you might be asking, is the relevance of all this? Does it matter whether or not Earth's life-forms control the planet's climate like a giant thermostat? The answer to this question will become more apparent as we go on.

Greenhouse Earth – The World One Hundred Million Years Ago

A hundred million years ago there were no Polar ice sheets; Antarctica was completely dry land and the Arctic Circle was an ocean of liquid water. This was a greenhouse world. Scientists can recreate the image of this world with confidence because there is much more evidence for the last 100 million years than for any time previous to this. So we now have the opportunity to see how the climate system acts in a world with different characteristics.

A hundred million years ago, Earth's tectonic plates were in different positions than they are today and the landmasses we currently know may have been up to several thousand kilometres further away. There was also a shortage of dry land around this time, with many of our present continents broken by seas and channels. North America, for example, would have been impossible to cross east to west without a boat or a pair of wings; Europe was little more than a collection of small islands.

This lack of dry land was due to an extraordinary high sea level during this period - 200 metres higher than it is today. Though unusual to us, these seas and oceans were around for millions of years, and dead sea-life was consequently deposited over an ocean floor that is now surface land. Today you can find limestone deposits in places around the world; this limestone is dead sea-life, and the area it is found in was once as much ocean floor as is the bottom of the Atlantic is today. For this reason, geologists refer to this period in time as the Cretaceous – meaning 'abundance of chalk'.

As we study the Earth 100Ma we are looking at the planet as it was during its last greenhouse period, when no permanent glacial was to be found anywhere on the Earth. This period didn't arrive suddenly, and probably took millions upon millions of years to arrive. Bearing in mind that there was no Polar ice in the Arctic or Antarctic, faced with such a dramatic lack of dry land we would expect that life had once been quite abundant in these areas. The evidence supporting this is very strong. This was an extraordinary period for our planet; a period where the Arctic Circle could have been home to warm-loving animals such as crocodiles and turtles, whilst dinosaurs could be found the Antarctic.

How High is High?

Throughout history, the oceans have risen and fallen against the continental margins. These changes can be several hundred metres in height, such as 100Ma when the sea was 200 metres higher then present levels. When changes of several hundred metres are considered in relation to the average depth of all the world's oceans (more than 4000 metres) they seem very insignificant, but they are known to have very strong effects on the climate in some regions.

The reason for this is complex but it mostly comes down to the shape of the continental plates and the gradient of their shelves. Generally speaking, most continental margins are flat until the point where they reach the deep ocean and the land slopes away rapidly, and often their changes in depth are just one metre for every kilometre moved across the continental shelf. This gradual incline towards the shore makes the matter of flooding all the worse, as the sea rises of several hundred metres then translate into several hundred kilometres of flooding inland. Evidence available in the sedimentary record can give fairly accurate ideas of sea level over the last 600 million years.

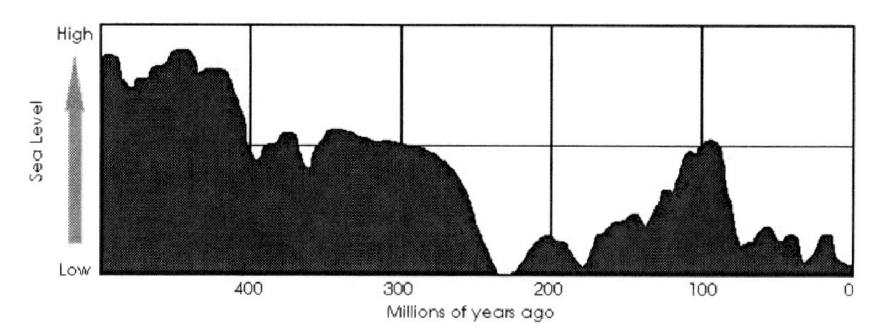

Figure 2.3 – Sea level over the last 600 million years has rarely been as low as it is today. We may think that the sea level a hundred million years ago was unusually high, but according to the record this height seems to be the average. We have built our present civilisation according to an extremely uncommon sea level.

The last time the sea was at a relatively high level was during the greenhouse world of 100Ma, but before this there was a period between about 550 and 300Ma when the sea level was at least the same height as 100Ma. For a period between 520 and 420Ma the sea level increased further to a peak of almost twice that of our Greenhouse Earth of 100Ma. One of the biggest lessons to take from all this data is that sea levels are rarely as low as they are in today's world, and that the sea level over the past 600 million years was about 150 metres higher on average than it is today. As a result, today we find ourselves fortunately blessed with an abundance of land.

The Effect of KT 65Ma and other Asteroid Impacts on Earth's Climate

Earth has always been hit by solar debris, both large and small, throughout its long history. Thankfully, today it suffers relatively little damage compared with what it sustained during the first 2000 million years of its life. Also fortunate is the fact that an inverse relationship exists between the sizes of these extraterrestrial objects and the frequency of their arrival. The smallest masses (with a diameter less than 10km) hit us far more often than the larger ones that usually hit on average every 50 to 100 million years. The last of these 'great' impacts occurred 65 million years ago and is known to science as the KT event.

Such large impacts are devastating to the planet; the soot and ash caused by the impact, and the inevitable fires that follow, fill the atmosphere and blot out the sun for years and years. Little sunlight disrupts the ability of life to sustain successful growth. As recently as 2001, researchers at the Californian Institute of Technology analysing helium isotopes from sediments left over from the KT impact confirmed that it had affected the Earth's climate for about ten thousand years. As a result, around 50 to 70 per cent of all the species on Earth became extinct.

The first changes to occur during an asteroid impact begin before the asteroid has reached the ground. The sudden emergence of a large sized object ripping through the atmosphere would have caused a shock wave travelling at the same speed as the moving asteroid – 20 kilometres per second. This shock wave would flatten trees and other objects for hundreds of miles around the impact site and heat the atmosphere as it moved. As the object impacted, there would be more consequences - the seismic waves caused by the impact travel through the interior of the planet and are similar to those caused by an earthquake measuring eleven on the Richter scale and up to 1000 times stronger than the strongest waves in recorded history.[iii]

Following the initial impact, the dust and soot floating above the impact zone, high in the atmosphere, would begin to spread around the planet, blocking most of the incoming solar radiation. The dust in the highest layers of the atmosphere would have taken many years to return to the surface. The subsequent cooling of the climate would have been rapid. It is also likely that the heat of the impact would have transformed much of the regular atmospheric nitrogen, oxygen and water vapour into nitric acid (a component of acid rain), which would have fallen to the surface and acidified the oceans. For life-forms, the outlook was already bleak.

Longer-term impacts involve the sudden injection of CO_2 into the atmosphere, caused by the burning of forests and plants. It would be a long time before the remaining land plants could prosper again, filling the void left by those wiped out, and beginning to pull back significant amounts of CO_2. Still, in relation to events of such magnitude over Earth's entire history, the KT event was a short, sharp jab – causing devastating damage but producing only short-term environmental consequences.

Cooling down: The Last 55 Million Years

As the world transitioned between the glacier-free world of 100Ma and today's ice-capped planet, something unusual occurred - causing a planet-wide warming that has rarely been seen before in the entire climate record. In 2003, the Ocean Drilling Project recovered sediment from 200 metres below the sea floor of an underwater mountain range in the north Pacific. It soon became apparent that this was no ordinary picture of climate when examinations of fossilised Foraminifera showed that a mass extinction had occurred around 55Ma in the deep and not-so-deep oceans. On land, the period is marked by a huge influx of Asian flora and fauna into Europe and North America, forcing many of their native species into extinction. The most unusual thing about these two changes is that they seem to have occurred over a very short period, somewhere between a few decades and a couple centuries.

A year after this discovery, scientists revealed that around the time these two things happened, an unbelievable amount of carbon had been forced into the atmosphere – between 1500 and 3000 gigatonnes (thousand million tonnes). The consequences of this sudden spike are vast. Over a period of only a few years or decades, atmospheric CO_2 concentrations rose from around 500ppm to around 2000ppm (for the entire span of recorded human history that figure had never topped 280ppm, until fairly recently).

The story begins in huge crater-like structures lying deep within the North Atlantic Ocean just off the coast of Norway. Here lay – and still lies – vast amounts of what is largely methane, a potent greenhouse gas. This is the home of clathrates, as it was 55 million years ago. Clathrates are compounds with cages of molecules that can trap gases such as methane. Today, methane clathrates may make up a significant portion of fossilized carbon, with the best-guesses ranging from 500 to 2000 gigatonnes (5-20 per cent of all carbon reserves).[iv] They exist mainly on continental shelves where water is cooler and where the suitable pressure and better supply of organic material is enough to keep the bacteria happy.

This provided the fuel. The ignition to the fuel came when magma managed to creep in nearby, providing all the impetus for a very big bang. As the clathrates heated up they expanded, and headed full-steam towards the surface. At the sea floor a huge explosion must have occurred, sending a great plume into the ocean. Most methane 'burned' with oxygen in the ocean waters, leaving CO_2 to arrive at the surface and bubble out into the atmosphere. With a lack of oxygen, and CO_2 turning the water acidic, life in the oceans must have been hell for most organisms, causing widespread extinction and a warming of the atmosphere leading to further changes in ecosystems on the land. The mean global temperature shot up by three to four degrees. It took 20,000 years for the planet to reabsorb this extra carbon; since then the Earth has cooled, returning to previous natural atmospheric CO_2 levels.

Scientists are now looking at the clathrates to explain other short-term explosions in global temperature found in the climate record. It is looking increasingly likely that methane clathrates are a hidden source of rapid global warming, waiting to pounce whenever tectonic conditions change... or whenever temperatures creep high enough.

It should be pointed out that the extinctions caused by this event are not great enough to begin a completely new Era – even though it seems a dramatic enough turning point in Earth's climate – because the planet was already a hot place. There were no ice caps, higher sea levels, less land surface, and all life was still recovering from the KT event of 65Ma, so its diversity and depth was relatively poor. The world had less to lose with such a large influx of CO_2 into the atmosphere; it had only a short distance to fall - not like today.

The cooling of the last 55 million years may be explained by the chemical weathering process we examined earlier, which depends heavily on the amount of surface rock available. For this to be the answer the last 55 million years must have seen an unusual amount of mountainous regions, and one mountain range turns out to be extremely unique in Earth's recent history. We are talking about the Himalayas of South East Asia, standing on average 5000 metres above sea level and crowning the Tibetan Plateau like a throne for the gods. This extraordinarily extensive mountain range is home to the highest *twenty-three* peaks on the entire planet – the tallest, Qomolangma Feng, reaching 8,848 metres into the sky. Covering more than 2 million square kilometres in area, the Himalayas and the Tibetan Plateau are certainly unusual.

The Himalayan Plateau has been forming, and growing, since the initial collision of the Indian tectonic plate and the Asian plate almost exactly 55 million years ago. This collision is gradual and ongoing. India didn't simply slam into Southern Asia overnight; it has taken every minute of that 55 million years to make the collision complete, and thanks to tectonic movement of the Earth's plates, India will continue to cram into Asia for many millions of years to come.

The large amount of sediment lying in the Indian Ocean, at the mouths of rivers coming directly or indirectly from the Himalayas also supports the chemical weathering theory. The majority of this sediment has been deposited within the last ten million years - almost ten times what was deposited 40Ma. It turns out that the Himalayas are so unique that they generate their own miniature weather system, including the South Asian Monsoon.

But there are often multiple factors pushing climate changes. The oceans can shift enormous amounts of warmth around on the surface of the planet. This is heat that the atmosphere alone cannot shift, and since oceans can reach every nook and cranny of the continental land surfaces, the potential to carry massive amounts of heat is quite large. Some experts point to a close correlation between major changes in climate and the changes of some major ocean 'gateways' through history. Gateways are narrow passages that link one ocean base to another and changes in gateway configuration affect the amount of water moved between each ocean, as well as the heat and salt carried by the water.

Over the last 55 million years there have been two major ocean gateway alterations: one at modern-day Panama and the second around the Antarctic. At the joint between Central and South America there was once an open passage between the Atlantic and the Pacific where heat and salinity could pass from one ocean to the other. This passage closed a little over four million years ago. Likewise, the Antarctic continent was once joined by the skin of its teeth to the tip of South America, and this may have helped to bring warm currents down from the north and keep the Antarctic glaciers at bay. When the connection was broken, a passage of sea – known as Drake's Passage – allowed water to flow all the way around Antarctica, and this kept warm currents away. However, scientists still question the strength of the gateway theories and it is thought that their influence is only to compound trends that already going on in the climate.

The driving processes that have caused Earth to cool over the last 55 million years continue even today and all the evidence so far seems to point to a continuing slow march onwards into more icehouse conditions in the future. But the temperature is not going to keep plummeting downwards like a falling dart. Tectonic processes and their feedbacks take millions upon millions of years before they are noticeable. In the meantime, there is nothing to stop short-term climate changes from increasing the temperature and melting some ice sheets.

Into the Big Blue

Not only do the oceans act as a large carbon sink, taking in substantial amounts of CO_2 from the atmosphere, but also their slow response times to climate variations help to add stability and cushion the impacts of rapid climate disturbances. A system of dominant ocean currents, hundreds of metres below the surface, also transports heat around the planet. The process driving the deep waters of the oceans around the globe is known as *thermohaline circulation* and results from variations in sea water density arising from changes in salinity and temperature.

An example of this can be found off the coast of Antarctica in the Weddell Sea. Sea-ice (ice shelves) forms at each pole during its own winter time. When ice forms in salty-water, like in the oceans, the salt is systematically separated from the water molecules and is expelled into the liquid water nearby. For this reason, the water immediately below sea-ice is always proportionally saltier than usual. With the extremely low temperatures of the poles the water is already very cold, and the saline water sinks downwards in a great column, reaching the ocean floor and spreading. Although this explanation is rather simplistic, this is essentially what happens.

The ocean floor is usually relatively hot, due to geothermal energy, causing the water to warm as it sinks. However, in the Northern Atlantic, the story is very different. The conditions at this part of the world are quite unique and form a conveyor belt, driving the entire planet's deep-water circulation, which is known as the Great Ocean Conveyor (GOC). Developed by Wallace Broecker at the Lamont Doherty Laboratory, the GOC model has transformed the way scientists now consider the nature of climate change.

Beginning in the Gulf of Mexico, where warm winds gather from further South at the Equator, we can construct a step by step analysis to the GOC model, and in particular the significant North Atlantic dynamo driving the whole thing. The warmer Gulf air begins to increase the temperature of the Atlantic water as it pushes it Northwards up the Eastern seaboard of the US and then Eastwards towards Northern Europe. This is known as the Gulf Stream and is the reason why places like Paris experience similar temperatures to New York, despite being 10 degrees of latitude farther north.

By the time the ocean water has travelled northwards, towards the base of Iceland, it has warmed significantly enough. Here it experiences a major drop in temperature as polar conditions and freezing Easterly winds from Canada rapidly cool it down. Nearly a quarter of all heat received by the North Atlantic from the Sun is transferred at this point from sea to air.[v] As a result the Gulf Stream winds warm and go on to heat Europe - whilst the water is cooled to the point where it begins to gain density (salt) and thus sinks.

This water is then pushed (or dragged) southwards at a great depth until it reaches the Southern tip of Africa. There it turns north again into the Indian Ocean where it once again warms and feeds back into the Southern Atlantic cycle. From here, it either returns to the Gulf of Mexico, or heads east into the Pacific. The North Atlantic Deep Water (NADW) conveyor belt drives the whole thing. It is the push-and-pull mechanism of all the ocean currents on our planet - although there are other, less significant ones at work elsewhere.

Climatologists first realised the influence of the NADW conveyor over the climate of the Northern Hemisphere while they were taking some of the initial Greenland ice core samples for studying CO_2. One of these ice cores revealed contained the pollen of a flower that only ever grew in extremely cold conditions. The scientists were confused – clearly this pollen had blown over from Canada some 8,000 years ago but climate evidence was suggesting that the world was in a state of gradual warming as it came out of the last Ice Age. Canada, therefore, should have been too warm for this flower to grow.

The bubbles found in this same ice core revealed something else. They showed that there was a catastrophic drop in temperature around 8,700 years ago and evidence elsewhere was about to reveal exactly why. During the last Ice Age, Lake Agassiz in Canada had almost completely frozen over and was being held in place by a glacier across the river valley to the East. This large body of water sat where today one would find Lake Superior and its companions, far up the St. Lawrence River valley. As the world began to leave the ice age and warm up, the glacier holding back the large body of water of Agassiz began to lose its hold and melted over a timescale that could be as short as a single year. Lake Agassiz began to drain away and millions of gallons of fresh water now raced towards the North Atlantic at a fantastic rate, via either the Hudson Straight or the St. Lawrence River valley.

As this fresh water arrived in the Labrador Sea it disrupted the NADW conveyor and rapidly reduced the salinity of the ocean, therefore slowing the conveyor belt. As a result, far less warmth reached the Greenland Ice Sheet and consequently the ice sheet began to grow. Climate scientists now realised that the mechanism of the oceanic currents was the salinity of the North Atlantic, and therefore a *saltier* Gulf Stream would do the opposite and *warm* the world.

The consequence of the melting of Lake Agassiz 8,700 years ago was another miniature ice age and a growth in the Northern ice sheets across the borders of Canada and Europe. What is most astonishing about the Lake Agassiz event is the speed of which the world changed – the disruption of the North Atlantic conveyor sent the world into Ice Age conditions within a period of just 70 years, a single human lifetime. Evidence does suggest that the climate system has not always been so susceptible to the salinity of the North Atlantic, and certainly not before the Atlantic Ocean had properly taken shape. But this certainly seems to be a primary driver behind our present climate system and the Lake Agassiz incident is a firm example of what can happen if the thermohaline circulation is disrupted and, more importantly, the speed in which the change can come.

---

[i] Ruddiman, W. F. (2001). *Earth's Climate: Past and Future*. W. H. Freeman & Co., New York. Pg 87-88.

[ii] Ball, P. (1999). *H20: A Biography of Water*. Phoenix/Orion, London. Pg 21.

[iii] Alvarez, L. W., Alvarez, W., Asaro, F. & Michel, H. V. (1980). 'Extraterrestrial Cause for the Cretaceous Tertiary Extinction,' *Science*, 208: Pg 1095-1108.

[iv] Schmidt, G. (2006). 'Methane: A Scientific Journey for Obscurity to Climate Super-Stardom,' http://www.giss.nasa.gov/research/features/methane/, Accessed 23rd March 2007.

[v] Broecker, W. S. & Denton, G. H. (1990). 'The Role of Ocean-Atmospheric Reorganisations in Glacial Cycles.' *Quaternary Science Reviews 305*. Pg 41.

Chapter 3 – The Spinning Orb: The Sun, Ice, and Everything In-between

Whilst atmospheric carbon dioxide is the 'thermostat' behind Earth's global climate variations, a second force can also have an independent effect on the planet, and the source of this force is 153 million kilometres *outside* of Earth's atmosphere: the Sun. In fact, this 153 million km figure is itself key to the issue, for the Earth's orbit around the Sun is not always the same.

Climate scientists use several sources as indicators of past climates on Earth: tree rings, pollen, ocean sediments, ice core bubbles, and a few more. Ultimately, there comes a point where it is impossible to find evidence of climates because they are so far back in history. We can only speculate what Earth's climate was like in the first 2,000 million years or so of its existence and the further you get away from the present the patchier the climate record is. However, science can still piece together what *is* available to reveal something fundamental to our understanding of the climate. And that 'something' is all to do with inconsistency.

The Wibble and the Wobble

We can witness the short-term influence that the Sun has over our climate several times a year, during the four different seasons. Each season is a transition period; we can see this in trees and other plant life in parts of the world affected by cold winters and mild summers. During the winter, some trees are prone to shed their leaves as they go into a sort of comatose state, waiting for the weather to warm up again. The advent of spring sees the leaves return and the flowers and fruits bloom. Summer is the transition from the bloom of spring to the gradual decay of autumn when many trees begin to shed leaves and some animals begin forging their nests for the long hibernation over winter. All life on the planet is subject to seasonal changes, some more than others, and the vast majority could not survive successfully without them.

We all know that the seasons happen because the Earth is tilted rather than upright (at an angle of 23.5° to its orbital path around the Sun to be exact), and that it revolves around the Sun once every 365¼ days (or thereabouts), travelling at a speed of 100,000 kilometres an hour.

The planet's rotation is what gives us day and night – if we didn't have it we would have six months of bright sunlight followed by six months of terminal darkness. In fact, because the planet rotates at the speed it does, the western edge of the Pacific Ocean is about a foot and a half higher than its eastern edge; this is caused by water being pushed to the rear of the direction of rotation, like a roller coaster pushing you to the back of your seat when it goes very fast. Rotation therefore plays a key role in shaping our ecological and geographical heritage.

So our tilt and orbit give us the seasons and the rotation gives us day and night, and we can depend on these two things to be immovable constants in the whole lifetime of the planet. Or can we? Well, as it turns out, Earth's movement through space is a lot messier than it first appears. For a start, the Earth is constantly changing its distance from the Sun, ranging from 153 million kilometres to 158 million kilometres, due to an elliptical - not circular - orbit. Though this elliptical orbit has only a very tiny influence over world climate on an annual timescale, over the long-term the influence is much greater. This is down to the fact that the *shape* of this elliptical orbit is also changing, sometimes becoming quite elongated, and sometimes almost becoming circular. So not only is the orbit never circular, but its shape is also changing from near-circular to very ovular, and back again, as time passes. Here the physics and maths can get quite tiresome, so we'll just say that the shape of Earth's elliptical orbit alternates between maximum and minimum every 100,000 years or so.

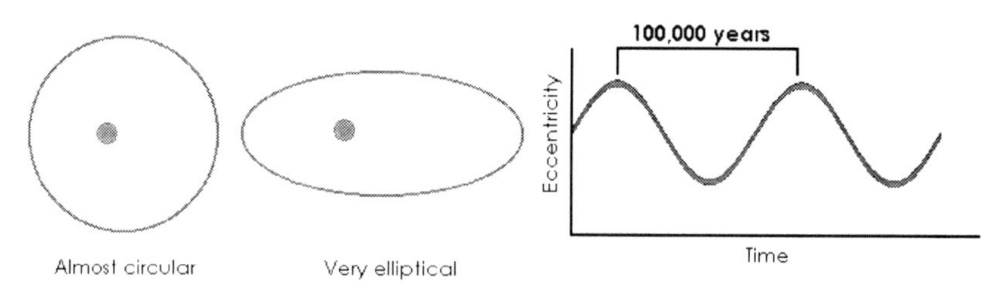

Figure 3.1 – Earth's oval-shaped orbit changes shape all the time. It alternates between ovular and circular in a gradual cycle of roughly 100,000 years. This affects how much sunlight the planet receives.

Without a tilted axis, Earth would have no seasons. A world like this would have bred an evolutionary tree extremely different from what we have, with creatures adapted to prolonged periods of dark and light. Fortunately we don't have either of these tilt conditions, but the slight angle of tilt we do enjoy is not always the same – it can be as low as 22.2° and as high as 24.5°, and it is always shifting between angles within this range. At present the angle is 23.5° and is decreasing towards the lower end of the range.

Because the tilt shifts between a low angle and a higher angle it creates a cyclical motion, which astronomers have worked out as occurring every 41,000 years. Over long periods this motion makes a kind of 'wibble' through space. The last maximum angle it reached, before shifting back, was about 24.1°, and that happened around 16,000 years ago. The last time the Earth was closest to being upright was around 36,500 years ago, when it managed to reach an angle of about 22.2°.

Though these angles seem insignificant the changes can be quite profound, either amplifying or suppressing the seasons on Earth - with the Polar Regions being the areas experiencing these changes the most. Obviously, the more tilted the Earth is the more extreme the seasons are. The effect is shown in Figure 3.2 below.

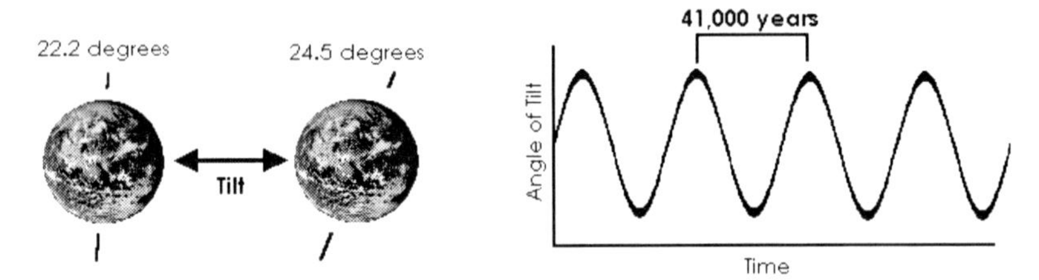

Figure 3.2 – Earth also changes its angle of tilt slightly over time, shifting between a 22.2 degree tilt and a 24.5 degree tilt every 41,000 years.

So, at one extreme, Earth's orbit will be very long and stretched out, and 100,000 years later it will be nearly circular. At the same time, the planet can be tilted far away from the Sun, and then 41,000 years later it will be tilted away but to a smaller degree. For the third important motion, let us imagine a child's spinning top. A decent spinning top can show us how the Earth moves around just as well as any book. First of all, the top spins very quickly around and around, the same as Earth's daily rotation. Secondly, it usually travels in a wide circular shape across the floor, which is similar to the planet's orbit around the Sun. But instead of 'wibbling' between one angle and another, the spinning top shows us a different motion: the wobble. The top's axis moves around in tiny circles, sometimes leaning inwards, sometimes leaning outwards. The Earth uses a similar motion in addition to its wibble, except – unlike a top – it wobbles after several thousand revolutions around the sun (Figure 3.3), rather than several times per revolution. One wobble is completed every 25,700 years, but this is also combined with a fourth orbital change to make a much more powerful cycle.[i] Earth's changing elliptical orbit (Figure 3.1) also revolves (like spinning a hoop around the top of your finger) and when you put the two together a stronger 23,000-year cycle is produced.

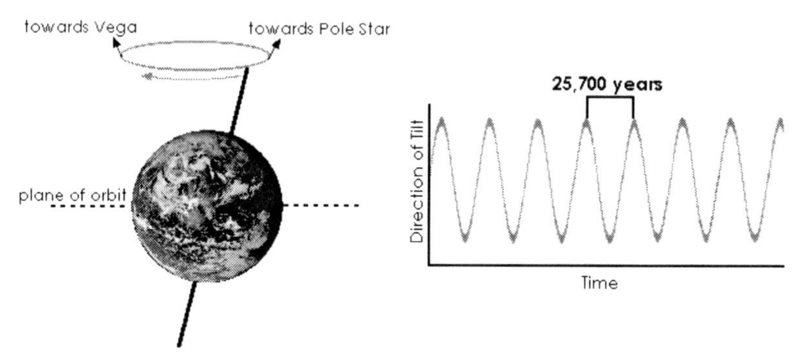

Figure 3.3 – The direction of Earth's tilt is also changing. It 'wobbles' around in a little circle, coming back to its original position every 25,700 years. At the same time this cycle is strengthened by the gradually changing motion of its elliptical orbit, and combines to make a stronger cycle of 23,000 years.

These three complicated changes (and a few smaller, less significant ones) have important effects down here on Earth's lowly surface. If we combine the three wavey motions of the above figures into a single wavey line the overall effect may be seen more clearly. What we can see from Figure 3.4 is that solar influence on climate is maximised roughly every 200,000 years.

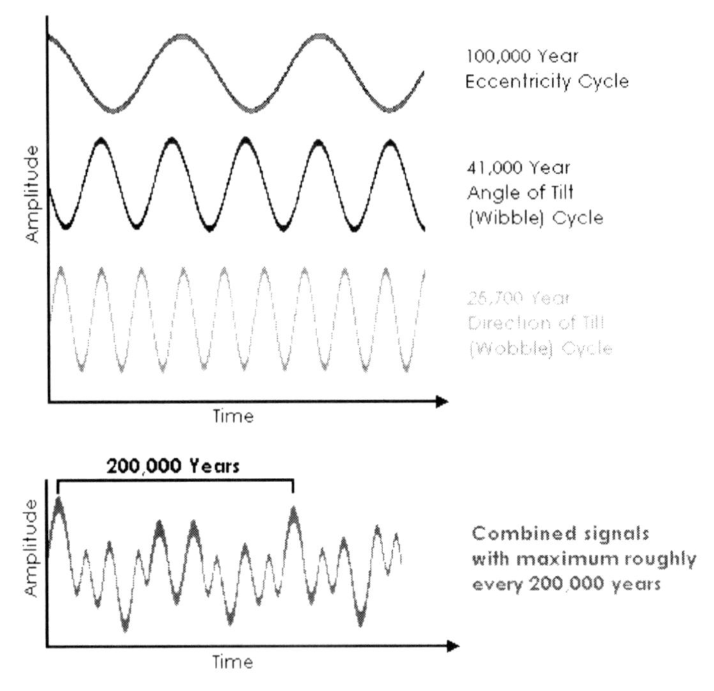

Figure 3.4 – Combining the three strong orbital cycles creates a variation where each individual cycle is hard to pick out. Together solar radiation is maximised around every 200,000 years, though this isn't always a certainty.

It was back in the first half of the nineteenth century when these strange behaviours were first discovered and calculated. Since then, science has managed to assess the past and possible future impacts of these wibbles and wobbles on Earth's climate. As it turns out, the wibbles and wobbles of our planet as it whizzes around the solar system have profound effects on its climate, with interesting consequences at the very extremes.

Wet and Windy

Monsoons are long lasting periods of immense rainfall, and usually occur in places which would otherwise be very hot, dry and sometimes arid. Indeed, monsoons provide so much fresh water to many millions of people that they are without doubt the most relevant example of a climate variable that we could use. Historically, populations living in monsoon areas are dependent on the life-giving rains that provide drinking and bathing water, as well as moisture for the soils in which they can grow crops. Though the advance of technological civilisation has breached the borders of many a monsoon country, especially India, the vast majority of people still depend heavily on the national agricultural output and monsoon failure could still potentially cripple the economy. Nowhere is this truer than the sub-Saharan Africa, in an area known as the Sahel.

If you're looking at a map of the African continent then the Sahel would be a band between the Sahara desert, to its north, and the tropical forests to the south. Sometimes the line separating desert and grass savannah is so distinct that it almost runs in a straight line from East to West. This part of the world is almost as sensitive to temperature changes as the Poles, and this is all down to its monsoons. Typically you wouldn't associate any part of Africa with monsoons, but Africa wouldn't be the land we know and love if it wasn't for them.

During the Northern Hemisphere summer the concentration of solar energy falls over the Atlantic Ocean to the west, and simultaneously creates a pocket of low pressure over west-central North Africa. The extra evaporation of ocean caused by more intense sunlight is drawn into this area of low pressure like a ball rolling into a divot. The moist winds are met by dry winds coming from the Mediterranean and Sahara to the north, and the two collide over the Sahel. At the peak of summer (August-September) this thick band of rain may reach to about 17° north[ii], as high as Lake Faguibine in the west to Khartoum in the east.

During winter the story is different. With the Sun over the Southern Hemisphere a high pressure cell develops over the north-west Sahara, pushing out winds to the south and forcing the rain-band to retreat to its southernmost extent during the peak of winter in February and March. The south-going winds are very dry, and unless you live along the Mediterranean coast at the very top, or along the 'Ivory Coast' to the central-west, you will see hardly a drop of rain all winter in North Africa. Vegetation follows the rain-belt, with the expansion of the tropical and savannah zones during summer and the expansion of the desert and semi-arid zones during winter.

This adds stresses to human populations during winter, and increases the need for a good growing season during summer to provide for the long, dry season ahead. This part of Africa is prone to famine because it is prone to drought during summer, which severely cripples the ability of the country to meet the needs of its population in terms of food. The famine of the early 1980s in Ethiopia occurred for a number of reasons, but the fact that there was a persistent lack of summer rainfall for several years running was certainly the main factor.

If the monsoon rain-belt is commanded by the input of solar radiation from season to season, surely it is reasonable to suspect that the longer-term orbital changes in radiation received by Earth should also affect it. Because monsoonal winters are dry, regardless of how much radiation is received from the Sun, orbital changes in winter radiation have absolutely no effect. However, because summer is the wet season the orbital changes do have an affect, and because summer is the *only* wet season, the amount of summer rainfall determines the entire annual average. So the changes to the Earth's wibble and wobble have the potential to wreak havoc on this particular part of the world because it will be affecting a very sensitive area, across a wide region, during a time of the year when both humans and plants are dependent on rains for survival.

Historical evidence supports the idea that the wibble and wobble influence Sahelian monsoons. Ten thousand years ago the amount of radiation received from the sun in North Africa was almost eight per cent higher than it is today, and when we look at past climate records we see the signs. In fact, many lakes in North Africa that existed 10,000 years ago are completely dry today, even during the summer monsoons. Furthermore, we can see that the 23,000 year cycle of radiation received by Earth is currently at its lowest point, and this tells us why there are fewer lakes in North Africa today, and why the whole region is so quick to become desert if it is pushed even slightly. With this in mind we can see that, as humans first developed basic agriculture around 10,000 years ago, the North of Africa was experiencing the peak of its monsoon activity. Since that time, with half of the 23,000-year cycle since passed, the region has been getting drier and drier with each passing millennia.

The important thing to remember is that Earth's orbit is extremely complicated and influences our climate, particularly monsoons. But climate changes on Earth only really get serious once they begin to alter the polar ice sheets, so is there any evidence to suggest that our orbit can do this? And that brings us nicely to one of the most important of all climate components, for both the past and for the near future: ice.

Ice

Did you know that only three per cent of Earth's water is fresh water? Since only a tiny amount (0.036%) of this fresh water is found in lakes and rivers you could say that virtually all of the planet's fresh water is locked away in the ice sheets.[iii] Nearly nine tenths of the planet's ice is sitting atop Antarctica, and the rest is mostly in Greenland. At the North Pole the ice is only around four and a half metres thick, whereas at the South Pole it is more than three kilometres thick. Antarctica's ice sheet is so huge and deep that it holds almost 9.7 million cubic kilometres of ice - if all of it melted it would raise every ocean and sea on the planet by more than 60 metres, as high as a moderately-sized office block.

As we've already observed, the last 55 million years has seen the planet cool down from its greenhouse conditions over 100 million years ago. Ice on Earth is itself a rather rare event, though there have been times when colossal ice ages have occurred, engulfing the vast majority of the surface into its freezing and harsh landscape. The last 40 million years saw the separation of Antarctica and South America, the joining of South America and North America, and the ramming of India into Asia to form the Himalayas, and all of these things helped to cool the planet down.

This long cooling period hasn't come at a smooth and steady rate, instead coming in fits and bursts of ice ages (both large and small) and warm periods too. Around 20,000 years ago the last small ice age (glaciation) was at its peak, and may have covered as much as a third of the planet's entire surface with frozen water. Since then we've had quite a run of luck, with the Holocene settling us into a period that may be quite unique in all of Earth's history – a time when both poles are covered in ice, when 14 per cent of the Earth's land surface is permafrost, and when we have ice in places like Kilimanjaro and New Zealand.

Glaciations are far from regular occurrences and start or end with just a small change in another climatic factor. For example, a period of high volcanic activity may provide enough heat to push back some of the hardier ice ages; similarly, a significant change in the configuration of the continents and the world may become more prone to glaciations. As we saw with the melting of the frozen North American lakes 8,730 years ago and the shutting down of the oceanic conveyor belt, glaciations are also prone to coming on very quickly.

So what controls the sizes of the polar ice sheets? Ice can only accumulate in temperatures that are consistently cold enough, so we currently see permanent ice sheets at high altitudes as well as high latitudes. We see glaciers at Kilimanjaro and the Himalayas because they are extremely high places, though their distance from the equator is only relatively small. Unfortunately, it is easier to shrink an ice sheet than it is to grow one. Ice forms from snow that has fallen, which builds and builds layer upon layer until it is compacted solid. Ice can accumulate at mean annual temperatures below ten degrees Celsius but never grow more than 0.5 metres per year at all temperatures. If temperatures were any higher than 10°C then moisture falls as rain and is unlikely to freeze, and if the temperatures were any lower than zero then the air would be very frigid and would carry little water vapour to fall as snow. In contrast, ice melting can begin at mean annual temperatures above -10°C and at 0°C the rate of melting can reach 3 metres per year, much higher than the rate of ice formation.

So all this really means that at temperatures below -10°C you're more likely to see ice accumulate than melt, but above -10°C the rate of melting begins to overtake the rate of accumulation. In a world currently obsessed with global warming, the consequences of this are quite significant, for higher temperatures will reduce the number and size of the pockets of air that will occur below -10°C. Enough warming over glaciers and ice sheets and you get a rate of ice accumulation that is worthless to the planet, simply because so much more ice is melting.

The early scientists who looked into how the Earth's wibbling orbit influenced the size of the ice sheets assumed that winter was the critical season of the year. It is quite a straightforward idea, and seems blindingly obvious to you and me: winter is when the snow falls, and when temperatures get chilly – surely this is related to the size of the ice sheets. However, summer is the really critical period for the ice sheets. Winter must be discounted for two reasons. For one thing, at high latitudes the sun is always low in the sky, regardless of the planet's wibble, and so the intensity of heat from the sun is never going to be large. Secondly, the high latitudes are always cold during winter, so ice sheets will always grow during this season. In contrast, summer at high latitudes can be much warmer than the winter, and it often breaks through the melting-point threshold if summer temperatures are particularly high. Because it is easier to melt an ice sheet than grow one, you could have a winter of widespread snowfall but any small increase in summer warmth and all this snow can be lost.

In the early twentieth century, the Serbian astronomer Milutin Milankovitch proposed that the Earth's tilt shaped the size of northern hemisphere ice sheets. He suggested that when the Earth is less tilted towards the sun, the summers in the northern hemisphere are cooler. This can trigger ice sheet growth that can cover much of mainland Europe, including the British Isles and Scandinavia. At the peak of ice sheet growth 20,000 years ago, the last time the northern ice sheets were bigger than they are today, ice forged its way down through Scotland and Scandinavia, the Hudson Bay and Boston - carving up the land as it spread, pushing mountains and hills aside, freezing lakes and drying rivers, and creating large areas of permafrost south of the ice, making hunting and gathering next to impossible for the human communities there. Milankovitch turned out to be right; the ice sheets depend heavily on the radiation they receive from the sun.

With the monsoons and now the ice sheets, a lot rests on just how hot our summers turn out to be. Next time you pick up a newspaper and it tells you that the years are getting hotter and hotter, resist the fantasies of sunbathing and strawberry picking; instead we have to remember how dry it will be in Africa and Asia, and what that will mean for millions of people who depend on seasonal rains. Also spare a thought for the break up of the ice sheets at the Poles - all that water has got to end up somewhere. Later chapters will reveal to us that hotter summers *are* occurring, and they are already causing problems just like these.

---

[i] Imbrie, J. & Imbrie, K. P. (1979). *Ice Ages: Solving the Mystery*. Enslow, Short Hills, N.J.

[ii] Tucker, C. J. & Nicholson, S. E. (1999). 'Variations in the size of the Sahara Desert from 1980 – 1997,' *Ambio* 28, 587-591.

[iii] 'On Thin Ice', *Scientific American*, Dec 2002. Pg 100-105.

PART 2 - People

From the great mass extinction of 65 million years ago a new order emerged. A world once dominated by reptiles had been eliminated, and another class of animal came to power: mammals. Scientists speculate that hominids evolved about 15 million years ago, originally from apes or similar primates. It is believed that just 3Ma, a species known as genus 'homo' walked the Earth – a type of mammal that was able to think and manipulate its surroundings a fraction more than other species; a mammal that was able to remember more information, and collect it within its brain, and not cease to learn in this way throughout its life. In time it would spawn a series of species that would dominate the Earth, and today one such species is at the top of every food chain, and can dominate or devastate any territory, ecosystem and habitat. For that species, survival is not dependent on the strength of food supply within its ecosystem because it exists across so many ecosystems, and has stepped out of the usual dynamics of nature in terms of population and distribution. In other words, humans are everywhere, and there are plenty of us.

We are the first living beings known to Earth that do not just exist for the purpose of existence, but instead we live for fulfilment of the mind – either through knowledge, happiness, love or power. We live for pleasure.

For hundreds of millennia, human beings lived as nature intended; our routine of hunting and gathering food, creating shelter and defending territory, was within complete harmony with nature. With time, the traditional hunter-gatherer would give way to the agriculturalist. This opened up a whole new world to our ancient ancestors, developing social complexities and enhancing critical tools of civilisation such as languages, written words and currency. Within a matter of mere centuries, and several early civilisations, the world had already seen the dawn of early monarchs, complex philosophies, long-distance trade, conquest and empire. But even these are just the first steps on the path to the world we now live in; in between is a long story punctuated by technological leaps and large social setbacks. But has there ever been an instance, in this long story, where climate has interfered (or aided) the progress of our species. Of course, the answer is yes.

Chapter 4 - Climate and Humans

How much does our species owe to the climate and how much do we suffer from it? To answer this it is best to go right back to the beginning - to a time before our species even came into existence – and see what the climate was doing at the time of our conception.

Did Climate Change Drive Human Evolution?

Evolution of life on this planet is often depicted as a large and extensive tree. Each major branch represents one of the five kingdoms of life,[1] and every one of these is divided into smaller branches – representing subdivisions known as Phylum. And each and every Phylum divides further into even more branches and so on and so on. Phyla are subdivided into Classes, which are subdivided into Orders etc. until each Family, Genus and Species is separate.

Every single species has its own unique branch – or twig - on this extraordinary tree, and the theory goes that every single one can be traced backwards to the same origin. The particular branch of the tree that led to *humans* has been created by five distinct developments:

1. Divergence from primitive apes (6Ma – 4Ma)
2. Gradual preference for bipedalism (walking on two legs) around 4Ma
3. Technology: use of stone 'tools' (around 2.5Ma)
4. Further divergence creating the genus Homo (2Ma)
5. Gradual development of larger brains (since 2Ma).

Anthropologists contest many theories about exactly what is the driver behind human evolution on Earth. Most likely it is the same thing that drives all – or most – evolution on the planet, but what is it?

One hypothesis suggests that the evolution of humanity is the result of a changing global climate. According to this theory, our evolution was most rapid at times when the climate was changing, forcing our early ancestors to

[1] The number is often disputed. Up until the 1980's there were only really 2 recognised Kingdoms of Life. Most experts today agree with the given figure of 5, although numbers as high as 24 have been suggested.

adapt; those that changed quickly and suitably were the ones to succeed. Out of all of our closely related ancestral species that diverged away from the apes, only we have survived to this day. From what we know of these close relatives, it is apparent that our success depended on our ability to adapt to many climates, foods and circumstances.

Widespread conversion of forestland into grassland around the world, between eight and six million years ago, seems to be the primary reason that human ancestors became bipedal - and hence developing use of their hands for crafting tools. The moment our ancestors started using their hands to manipulate their surroundings they were on the path to cave paintings, flint spear-heads, culture, speech and so on. Maybe we owe our existence to a climate shift that took place millions of years ago, favouring species that could leave a forest habitat and exist in grassland.

This climate hypothesis of evolutionary drive can also be applied to all life on Earth, not just humanoids. Long-term climate changes on local and global levels produce bursts in evolution, as species are forced into new directions. Alteration of habitats puts new demands on species that had already successfully conquered that habitat, and had established themselves in a balance within the ecosystem. The ultimate change in climate favours those species that had evolved traits most useful for survival in the *new* environment.

Ultimately, the fragmentary record of human remains is so poor (with samples scattered sparsely over the last few million years) that the degree to which climate has affected human evolution is unknown. However, some evidence can link glacial cycles to human evolutionary bursts, but this is also very rare. It is thought that truly modern humans, with very large brain cases, only appeared in the last 200,000 – 100,000 years, within a sequence of glacial cycles. The warmer and wetter parts of the cycles encouraged minor population explosions, and the cooler and drier parts of the cycles then whittled away those humans who were not best suited to surviving the severe environment. In this way, perhaps, it is possible that climate change created 'windows' of opportunity for those early humans to evolve. The problems caused by a huge lack in early human remains are many, and the record is populated by samples 100,000 years apart. Because of this it is extremely difficult to construct an accurate picture of human evolution. However, the fraction of data available does apparently show an evolutionary characteristic working in fits and bursts, similar to the working of the climate.

New Shores

The story of humanity is a 200,000 year long saga, beginning with no more than a few hundred people in Africa and ending with their 7,000 million descendents spread out all across the globe. In between is a story of survival, isolation, migration, conquest and imperialism (though much more of it than you might think occurred before we first began to write things down). How did these early humans spread so far and wide, and what compelled them to keep wandering from home, with every passing generation venturing further and further into uncharted territory? Before the mid-1980s, all scientists had to construct the picture of human migration were scattered bone fragments and artefacts, which left an awful lot to the imagination. It is only within the last twenty years that scientists have been able to unravel the genetic code in living humans to create the mother of all stories.

Every drop of blood contains DNA, a genetic timeline waiting to be read. The human genetic code is 99.9 per cent identical throughout the world.[i] The remainder (0.1 per cent) is what distinguishes everyone from everyone else (eye colour, hair colour, etc.). The genes of your parents are only two threads of a complete tapestry of genes that go into making you who you are. By comparing your genes to someone else's, geneticists can work out how related we are to each other. Similarly, by comparing genes of people all over the planet, geneticists can begin to map the story of migration of all the different communities and estimate when we parted ways over the years.

In the mid-1980s, geneticists used mitochondrial DNA (mtDNA, which is passed directly from mother to child) to compare women from around the world, and found that women of African descent showed twice as much genetic diversity than their sisters outside Africa. Geneticists now believe that all living humans can draw their family tree back to the same woman, living some 150,000 years ago in Africa - what they like to call a 'mitochondrial Eve'. Of course, this 'Eve' was not the only female human alive at the time, but it seems that only *her* descendants managed to multiply beyond their usual realm, and infiltrate every human community alive at that time, over the long term. The oldest known fossils of modern humans were found in Omo Kibish, Ethiopia, and this seems to further support the African origin of humanity.

It seems that around 70,000 – 50,000 years ago the descendants of 'Eve' managed to push their population further and further north, until they reached the Nile valley in North Eastern Africa; as few as one thousand people may have reached this point. The climate at this time was in the depths of an ice age, causing sea levels to fall and extra land to appear. For this reason, the early inhabitants of the Near East may have crossed the mouth of the Red Sea, between the Horn of Africa and Arabia, which would have been divided by only a few miles of water. If not, then it is probable that the Nile valley populations began to push into Palestine through the Sinai, and continued to march on (over many, many generations of course) through Persia, the Indian subcontinent and down into Australasia. Fossils at Lake Mungo in southern Australia (45,000 years old) and Malakunanja in northern Australia (50,000 years old) show that it only took around 20,000 years for human to move from Palestine to Australia, a rapid rate for a people with only a slow expansion of population forcing them to push on further.

Between 40,000 and 30,000 years ago humans seemed to reach Europe, with genetic similarities showing that Europeans originally came from India rather than straight from Africa. A cave in France has shown that these humans coexisted with the original inhabitants of Europe, the Neanderthals. Whether these two species watched each other suspiciously from afar, shared intelligence and commodities, or fought bloody battles, is still up in the air, since little convincing evidence exists. Whatever the situation, it is certain that Neanderthals were forced into smaller and smaller pockets of existence that eventually dribbled off into extinction. No Neanderthal mtDNA is found in modern humans.

During the last ice age, humanity experienced two invaluable changes: extra land, and speech. With the sea level decreasing as usual during an ice age, 'land bridges' began to form in vital places across the globe. When the sea level fell the land separating Asia from North America, called the Bering Strait, became physically habitable, and provided early tribes living in North East Asia with a means to travel across, then downwards, into a new continent. The Mongoloid traits of Native Americans seem to support the idea that America was populated from Asia. Eventually, given a great deal of time, human beings would migrate as far as possible and populate both North and South America.

By the time a land bridge appeared connecting the islands of South East Asia with the desert country of Australia, nearly every bit of available land on Earth was populated, at least in part, by our species. These early hunting and gathering communities lived in sparsely distributed groups, divided and restricted by the barriers of nature: ocean, mountain, and ice.

Humanity developed the ability to speak out of necessity, to help them with hunting and survival in harsh climates. Words were not only used to point out prey, or help build shelter and fire, but they began to be used as labels, naming everything in sight. Finally, around 15,000 BCE, the sea level rose again as the ice caps subsided. This rise caused massive changes in the way Homo sapiens began to exist. Great cut-off points developed around the globe that prevented little, if any, human interaction between certain communities. For example, the expansion of the Sahara Desert, which today engulfs Northern Africa, provided an isolating barrier between communities to the South and their Mediterranean counterparts. The land bridges and shallow sea leading like stepping stones from Asia to Australia became much harder to pass, virtually cutting off the intercontinental link. The Bering Strait once again became impassable; this is the reason why, several thousand years later, Christopher Columbus thought he'd discovered a new continent and was rather shocked to find people living there.

Additionally, rising sea levels may have devastated human populations around the globe. During the ice age, human communities had settled and established rudimentary villages, mostly in areas close to, or on, coastlines – probably because of the fishing opportunities and the possibility of evaporating sea water to produce salt for preserving meats. It is therefore very likely that many of the established settlements of ice age populations exist in areas now submerged under sea level. It makes sense to therefore presume that the end of the last ice age saw a generic demise of humankind, especially in terms of population.

Every single one of us alive today descended from early Africans. Wave after wave of genes pushed the human population north into Europe, south into Australia and east into the Americas. Humans began to develop the physical dissimilarities we see today, and the cultural and social differences, every time they became isolated. The human species has populated virtually the planet's entire land surface in waves of expanding and integrating populations. Though we currently number around 7,000 million individuals, and are divided in every way possible - be it political, cultural, theological, or financial - it is comforting to see that we all can trace our origins back to a single community, living as brothers and sisters more than 200,000 years ago.

The Holocene and the Dawn of Agriculture

One look at the history of the human race reveals one stark fact: the vast majority of human accomplishments have occurred within the very recent past. In terms of individual achievement, technological invention and material production, the trend is obvious: the closer human history gets to the present, the more rapid the human landscape changes. In other words, by the time you wake up tomorrow the world will have changed more overnight than it did within a single millennia 10,000 years ago. More inventions, more scientific breakthroughs and more exciting things happen the closer one gets to the present. Just as the 20th Century was jam-packed with excitement compared to the 10th Century, the last 1000 years has seen much more advancement than the thousand years before that. The process is like a gathering snowball, and modern experts enjoy predicting the future with this runaway process in mind.

History gives us two good milestones of human advancement since the beginning of our species. First, human societies only started recording their goings-on about 5000 years ago, and this is a point in time many people regard as the beginning of 'modern' history. But a second, and more suitable, milestone exists around 10,000 years ago – at the first appearance of agriculture. The emergence of farming, in a world of the traditional hunter-gatherer mode of human existence, seems to be a trigger for 10,000 years of acceleration into what we see today. Agriculture was so important to human development because it was a break from nature. What experts have dubbed the Neolithic Revolution was the first time that the initial hunting and gathering lifestyle of the average human being was discarded in favour of something even slightly artificial. Farming began in the Near East, in a region known as the 'Fertile Crescent' - the arc-shaped strip from Palestine in the west to Mesopotamia in the East. This is not surprising; the area was already unusually rich in large-grained cereals, which, of course, humans find incredibly ideal for transforming into food. The fact that the Near East was probably the first port of call for our species as we left Africa also gives the region a head-start over the rest.

Although the reasons why these communities initially began farming are a bit of a mystery, the fact that they did appears to be inevitable. By 10,000 ago the land bridge between Asia and America had disappeared and the people of the Americas were living in complete isolation from the rest of the world. As farming took off in the Near East during the ninth millennium BCE - spreading west into Europe, east into the Orient, and south into Africa - the Americas were left out. However, farming managed to emerge in the Americas around the fourth millennium BCE with no sign of influence from the rest of the world. The fact that farming appeared in two separate places, completely independent from each other, suggests that its emergence was not sheer luck. Instead it seems inevitable that humans would one day discover farming.

Agriculture reduced nomadic wanderers to a minority and where the best farming land lay, large communities began to settle and build. Eventually these villages would become great cities and lead to bigger populations, nationalism, armies, fortifications and war. And thereafter the ball kept rolling, creating our history.

But we have to ask ourselves why such a colourful history of great civilisations, empires, religion, conquest, space exploration, medicine, skyscrapers and so forth, is packed into the last 10,000 years, when clearly human beings have been around for more than 200,000 years. Yes, the answer lies in the climate.

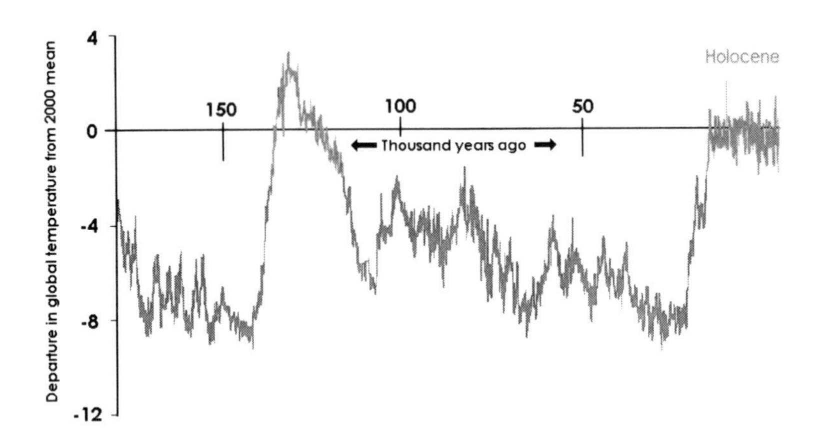

Figure 4.1 – The generally tranquil temperatures of the last ten thousand years (known as the Holocene) is very uncommon compared with the rest of the climate record, particularly the 200,000 years preceding it.

Figure 4.1 shows data from a Greenland ice core, which can be translated to show global temperature change in the last 200,000 years. At the right hand side of the graph (the last 10,000 years) is a distinct period known to us as the *Holocene,* and is unusual for two reasons. First, it is unusually warm: before the Holocene, the last time such a warm temperature was reached was 120,000 years ago. Second, and more significantly, the Holocene is unusually stable: no period in the last 200,000 years has seen such climatic stability lasting for an interval of 10,000 years. Earth's thermostat seems to have been set at a comfortable 14°C over this entire period, making it very suitable for a species such as ours to advance both technologically and socially.

Since human civilisation is founded on agriculture, it would make sense that agriculture and civilisation could only establish themselves and be developed during a period of persistent warmth and stability. All of our history seems to fit snugly into this little pocket of warmth that is the Holocene. As Michael Cook observes in 'A Brief History of the Human Race':

The Holocene, then, was the window of opportunity for the making of history. If this is right, there is no reason to ask why humans should have waited so long before making history; it seems that they jumped through the window just about as soon as it opened for them[ii].

Luckily, those humans that did jump through this window were ready and waiting; already with the intellectual capacity and potential, the human race was as ready as it ever would be to move on to the next level. Most probably, if humans had still been without language and culture, the Holocene window would not have presented a new opportunity until many thousands of years later, when these milestones were reached. It therefore seems to take a little more than just climate change to drive human civilisation, but that is not to say that climate change was not key.

Why Everything is the Way it is Today

In 2003 the United States led a military force into the Fertile Crescent to seize control of Iraq. The US is today the single most powerful nation on the planet, spending more on its military than the next ten powerful countries put together. They dominate the world both economically and politically, and things have been this way for much of the twentieth century. Why is this? We could just as easily substitute in the words 'United Kingdom' in place of 'United States' when we talk about the *nineteenth* century. But why was this?

Iraq, on the other hand, was once home to the Mesopotamians – arguably the world's first civilisation, but today the country is deeply impoverished. When we look at the great empires of history, we have an abundance to choose from in the same part of the world, beginning with the Mesopotamia: the Egyptians; the Persians; the Greeks; the Romans; and much later the Spanish and the British. There are many other classical powers that we could mention: the Assyrians, the Minoans or the Sumerians. All had developed trade routes, philosophies, governments, social classes and technologies. This special part of the world is, of course, the Near East and the land surrounding the Mediterranean Sea. But why *here*?

At first it seems strange that we would see such a flurry of activity, which would eventually dominate world affairs, occurring outside of Africa, where humans originated. In fact, it would be reasonable to presume that if the hallmarks of great civilisation would emerge anywhere it would be the place where humans have been established the longest – where roots and traditions had been more firmly placed, ready to be the foundations for something much greater to come. If it has something to do with the emergence of farming (which we recall occurred first in the Near East) then why do we not see the same flurry of imperial activity take place in the Americas, where we know farming also developed independently?

In order to answer all of the above questions, let us pose two more by looking at the 2003 US-led invasion of Iraq. First, why was North America the offspring of *European* civilisation and not one of the Central American powers such as the Aztecs? Secondly, how did a nation, that was up until the middle of the last millennia nothing more than a collection of indigenous tribes, lacking anything close to a civilisation, manage to invade land that once belonged to the world's first civilisation? In other words, why is today's world dominated by Europe and North America, and not any other viable alternative in Asia, Australia, South America, or even the Near East?

The way that history has happened may seem to be the result of one giant race between peoples for dominance – a race that Europeans began to lead several hundred years ago and eventually won (and thus went on the dominate everybody else). Obviously, to a large degree this is absolutely true, but if we look deeper we begin to see that things might not have been equal from the onset. Mediterranean and Near Eastern tribes had a distinct advantage over the rest because they were lucky enough to exist *where* they did at the dawning of the Holocene. This was the right place to be.

For a start, Europe and the Near East are located excellently in geographical terms, and this was essential for the region to be the first to develop agriculture. Australia by contrast is incredibly isolated and never independently created agriculture. Before the first European settlers arrived in the eighteenth century, the hunter-gatherer cultures there were very much the same as they had been for thousands of years. The fact that Australia is isolated from the rest of the world by sea certainly prevented agriculture from being imported sooner than it was, but what prevented agriculture from being developed by Australians *themselves* is to do with the land. Australia is incredibly flat, relatively speaking, and a lack of hills and mountains seriously decreases soil formation, and hence fertility. It is also incredibly arid, and what rainfall there is can prove unreliable and limited to small regions.

Critical innovations, such as domesticated plants, spread faster within each climate zone[2] than between them. This matters enormously when we look at the Americas. In both North and South America, climate bands are much smaller than they are on the landmass of Europe and Asia. For farming to emerge, and then spread, in the Americas would have been incredibly difficult because of these bands. Whilst we know that farming did eventually develop here, the important thing is that it took a lot longer than the Near Eastern equivalent. A critical reason for this is that there were fewer suitable plants to domesticate. Eventually maize and beans became predominant farming crops and spread from Central America northwards into the present-day US, but the spread was slow and not always successful.

The presence of the huge Amazon rainforest to the north east of South America squashes other climate bands up into thin strips along the western edge, where the Andean mountain chain runs for several thousand kilometres. This limits things for the inhabitants, as does the fact that both the North and South are equally misshapen, with lots of land at the top receding down to narrow points at their bottom. In other words, geographically there isn't much room to get a civilisation up and running, and where you *can* it is very hard work exporting it to other lands. The climate zones get in the way more than they help. This is also why we often get stuck listing great American civilisations once we've thought of the Incas and the Aztecs. And remember, the Aztecs of Central America were perhaps the continents most advanced, yet they lagged several hundred years behind their European rivals when the Spanish ships first landed.

Africa is also very much a victim of its climate. In the north is the Sahara desert, which isolated African tribes from Eurasia for thousands of years until fairly recently in history. Without that desert, Africa may have shared richly in the wealth of the European expansion, rather than becoming a colonised victim of it – especially because it is physically connected to the Near East by land in the north east (Egypt), and virtually connected to Spain in the north west (the Strait of Gibraltar is certainly crossable, even for ancient peoples). Rainforest in the centre of Africa also limited its progress, but what really mattered was the aridity of the north and the Sahel region. The large climate blocks of the rainforest and the Sahara desert made it difficult for movement

[2] A climate zone is an area of the Earth's surface that experiences a particular type of climate. Tropical forests and grassland savannahs are just two examples.

of ideas - especially religion, language and trade, as well as just farming. Besides, for many African communities, the hunter-gatherer lifestyle was well suited and working fine – farming may have seemed (to some) like a lot of effort for very little worth.

One African community that discovered farming fairly early on were the Egyptians, but this is purely because they were lucky enough to live next door to the Fertile Crescent of the Near East and because they had the River Nile. Flooding of the Nile makes the banks on either side extremely fertile and fantastic for agriculture – something quite unique in Northern Africa. The Egyptians flourished once they'd discovered the benefits of civilisation but they became part of the Near Eastern world rather than Africa. Egyptian civilisation never seemed to pass with any substance into the rest of the continent, with only one other community – the Nubians – adopting some of their ways.

For Asia the story is quite different. Powerful and advanced civilisations took off here quite early on, and its agriculture was able to spread quite freely around the continent's many river valleys once it had been imported from the Near East. Unfortunately, Asia is almost completely penned in by the mighty Himalayan Mountains to the West, and this served to keep the thriving ancient civilisations of Europe and Asia apart. There was trade and cultural crossover occurring between the two, but for the most part there was little real spread of civilisations going on. Though advanced and militarily-minded, ancient Europeans never really ventured into the east to expand their influence – Alexander the Great tried it around 329 BCE with a journey around the arid Hindu Kush and then down the Indus Valley into India, but an overstretched military held him back.

Asian civilisation seemed to lack enthusiasm for leaving Asia, with only the Mongols really taking the opportunity to venture west. Even in the south, where the passage between China and India is more agreeable, exchange of cultures was minimal, since both had well-established and complex civilisations already at work, and had little need for importing someone else's. Few major ideas seem to have spread from one to the other - the biggest being Buddhism, which passed from India into Tibet, through China, on into Korea and as far as Japan.

Near Eastern farming remained a regional phenomenon for several centuries before it appeared on other continents. However, once civilisation came to the Mediterranean it really took off in a big way, advancing well beyond the Near East. Again, what really held back the Near East was its climate. The Fertile Crescent may have been the cradle for agriculture and civilisation but it is, after all, only a small zone of land in a region that is largely arid and difficult. Were it not for the south Asian monsoon sucking all the moisture away from the Near East and dropping it over the Indian subcontinent, the picture would be quite different. In fact, it is interesting to note that modern day Arabia – which is more than 95 per cent desert – would be tropical jungle, were it not for the prevailing winds pushing the Asian monsoon away and to the east. Imagine how different the world would have been if a powerful civilisation could have developed, and been sustained, in this area - especially when you consider that today the region sits atop the largest reserve of oil in the world.

Mediterranean civilisation was able to expand north and west into Europe, where there was plenty of temperate land on which to develop. There were few climatic limitations holding them back and few geographical features – no deserts and no Himalayas. Europe is also unusual in two other respects, and both are linked to climate. Firstly, Northern Europe receives a high volume of moisture from the Atlantic Ocean, filling the land with rivers and fertile soil. Secondly, with sea levels as they are at present, Europe has an abundance of coastline – which, as we've already seen, provides early human groups with plenty of fishing opportunities, as well as salt for keeping meat. A mixture of imperial ambitions, trade and healthy competition helped drive the intellectual and technological achievements of European civilisations. When the time came to invade the New World across the Atlantic Ocean, the rest of the world was lagging behind considerably.

Today's United States of America is founded on a European concoction of tradition, philosophy, government and religion, which owes a lot to the conquests of the British, Spanish and Dutch. American language is largely English, though the Spanish colonialism that occurred in Central America means that there are also a lot of Spanish speaking people in the US. American government is even loosely based on the ancient Greek/Roman senate system, and its predominant religions are also (like most of the world) Near Eastern exports. When capitalism came to Europe and the US, these countries were well equipped to make the most of it, whilst others, such as countries in Africa and South America, would quickly become its victims.

History thus owes a lot to the climate. Were it not for the environmental circumstances facing the inhabitants of the Americas, Asia, Australia and Africa, civilisations from these continents could have easily become as dominant in the world as the European family (the US and Canada included) is today. Certainly these regions would have eventually created civilisations both advanced and imperial-minded, given time, but Europe happened to do it first. What this proves is that the climate has always been, and will always be, the natural phenomenon that defines our species.

---

[i] Shreeve, J. 'The Greatest Journey', *National Geographic*, March 2006.

[ii] Cook, M. (2004). *A Brief History of the Human Race*. Granta Publications, London. Pg 7.

Chapter 5 – Under the Influence: Climate and Human History

As sea levels rose the land became more fertile, allowing more and more food to be grown, particularly along the lowlands and valleys of the Nile, Tigris, Euphrates, and Indus. It is possible that the rise of civilisations in these areas at this time - around the third millennium BCE – was something of a necessity if there were growing numbers of people migrating away from the increasingly desert-like Gobi, Sinkiang, Rajastan, Afghanistan and Arabia.

Arguably the first great and glorious civilisation was that of the Egyptians, who built the grandest monuments and the world's largest temples – in fact, the Temple of Karnak just outside present day Luxor is the largest temple anyone has *ever* built. But the Egyptians were great only because of one thing: the River Nile.

The Nile is the only great river on the planet that runs from South to North on the map, and makes its way from Lake Maputo about two degrees above the equator, to the Mediterranean Sea about 32 degrees above the equator. Because it runs through a highly desert region it acts as the only useful fresh water source for hundreds or even thousands of miles. Nile flooding was (and still is) vital to the Egyptians, since the majority of the country is useless desert, but flooding was irregular and subject to seasonal change, causing many problems for inhabitants. To tackle this problem the Egyptians developed a system of irrigation that enabled them to have regular fresh water supplies virtually all year round. This leap in technological thought was soon to be accompanied by the rise of mathematics and science to successfully construct water features like drainage and waterwheels etc. Calendars were devised from the annual fluctuations of the Nile's water level and flood rate, and the pharaohs were soon demanding more complex bureaucracy and taxation. All these inventions came about thanks to water management.

Interestingly, the first canal linking the Red Sea to the Mediterranean was built at Suez in the twentieth century BCE, at the command of Pharaoh Sesostris I, around the time when the world's sea level was at its highest for probably a few thousand years. Perhaps the higher sea level made the idea feasible, and the reconstruction of the canal under management of Rameses II (reigned 1304 – 1237 BCE) several centuries later, came at a time when this prolonged warm period was coming to a casual end. The early Israelites' escape from Egypt around 1230 BCE may well have been the result of a fluctuation of the Red Sea waters (such as a storm surge) over the shallow ground during a period where the sea level was higher that it is today.

In his book 'Climate, History and the Modern World' Professor H. H. Lamb suggests that there may even have been a link between times of climate stress and the birth of the mainstream religions we see today.[i] According to him the "enthusiastic missionaries and/or armed supporters" of a religion may have found social and political conditions favourable during a period where globally increased variability of the weather had led to a breakdown in the traditional way of life and its ordered customs. "There is evidence of such breakdown, through drought, in parts of the Mediterranean world about the time of the spread of Islam in the first millennium after Christ."

So too, perhaps, did climate shifts in Asia spawn the rise of two other great philosophical movements; the stress caused by a changing climate around 600 – 200 BCE may have inspired the teachings of Buddha (563 – 483 BCE) and Confucius (551 – 479 BCE) who each tried to solve the problems of human suffering.

Mycenae

The ancient civilisation of Mycenae was one of great power and immense wealth. Described by Homer as "the Golden City of Heroes", Mycenae was based to the southern part of mainland Greece, close to Athens. It was due to this advantageous positioning that the might of Mycenae could run most of Mediterranean shipping and fishing during 1500 BCE as well as defeat the Minoans and invade Troy twice. Unfortunately the Mycenaean people happened to exist at a time, and in a place, which was about to get more than its fair share of bad weather. This would have disastrous consequences for their empire and it all seemed to crumble overnight.

A drop in temperature and a growing ice cap around 1200 BCE caused the polar front to weaken, forcing westerly Mediterranean winds to push further north. Usually this storm chain brought wet summers to South Eastern Europe and carried its rain on down into the eastern Mediterranean. But because of these Mediterranean winds the storm clouds stalled, and most of the precipitation fell over the Hungarian plains. So the only rain over the sea was brought in by the Mediterranean weather system. But this one came down low, bumping up against the mountains of Western Greece and dumping all its rain there; so for almost a century there was virtually no rain falling between this area and the Black Sea – exactly where Mycenae was located. Collapse followed the change and thereafter began the Greek Dark Ages – the doldrums of transition between Mycenaean Greece and the rise of the Greek Empire.

Biblical Flooding and Other Impacts

Almost every school child in the Western world will be told the story of a man named Noah and a great worldwide flood. This is an Old Testament story, telling of God punishing the human race for its sins by killing virtually every living thing on Earth, not including the good man Noah, his family, and two of every living species. Can flood legends, handed down the generations over thousands of years, be explained as an act of nature rather than divine intervention? After all, the Babylonians had a similar tale about Gilgamesh and other Old World cultures also have tales of widespread flooding, so there must be an explanation that satisfies all without bias.

As it turns out, the idea of a 'world-wide' flooding maybe the result of generations of exaggeration. The last time the Earth experienced massive flooding on a scale even close to *total* was 100Ma, which we already know was not even close (and at a time when there were no humans). So aside from the exaggeration, is there anything in recent history which can be interpreted as a rapid and devastating flood- one that could have been devastating to some communities and therefore passed on through the generations as such?

Two geophysicists, Ryan and Pitman, came up with one candidate back in 1998. They proposed that the Black Sea had rapidly increased in volume around 7600 years ago. According to their hypothesis - known simply as the Black Sea flood hypothesis – it was at this point in time when the last of the northern hemisphere ice sheets melted and caused the oceans to rise slightly. Whilst this made little difference elsewhere around the Northern Atlantic, in the Mediterranean – and specifically the Aegean, to the east of Greece – it was too much. The rising Aegean spilled into the Sea of Marmara, which at the time was a lake in Turkey. As the Marmara rose, it overflowed into a freshwater lake to the North East.

Ryan and Pitman calculated that this overflow of water from the Marmara into the Black Sea 'lake' would have raised the level of the lake by 15 centimetres in a single day. This is fast enough to advance the lake across a mile of low-lying land every day, continuously, transforming the lake into the Sea we have today. The pair also concluded that this relentless rise of the Black Sea, caused by overflow from the Aegean, would have displaced thousands of people in the areas surrounding the lake by flooding their settlements. These scattered peoples would have carried their tales of great flooding across the regions and gave birth to the many different flood legends of the Old World.

The best cases of climate changes affecting other civilisations throughout history are based in the Americas. Cultural records from arid regions show that a lack of water can devastate a strong civilisation just as much as too much water. Prehistoric cultures tended to expand across the continents during periods of plentiful rainfall, and then retreated back to the dependable sources of water (such as big rivers and lakes) when long lasting drought set in. Such fluctuating periods of wet and dry occurred in the Colorado Plateau in the American Southwest between 300CE to 1300CE. At this time, the Anasazi people abandoned their wonderful cave dwellings, cut into the cliff sides above the plateau, very abruptly - almost as if they all decided to pack up and leave at the same time. Why such a competent and successful people would abandon the entire region remained a mystery to scientists and historians, until studies of nearby tree rings revealed there had been widespread drought throughout the period, probably persisting for years until the Anasazi couldn't take it any longer. This wouldn't be the last time a successful civilisation would be wrecked by a nasty change in the climate.

The Rise and Fall of Rome and Petra

The ancient city of Petra was one of the most important in terms of trade during the period around 500BCE. Based in modern-day Jordan, the city of Petra was a jewel of its time. It survived down a deep and mysterious chasm in the Jordanian landscape thanks to technological genius and the creation of a water supply-line that brought the valuable substance down miles of tiny waterways carved into the rocks. The city experienced a surge in trade because it had the best possible position of all the civilisations - directly in the middle of European, North African and Near Eastern trading routes. A population of 30,000 people managed to live down a three hundred metre deep canyon and have all the luxuries of any other civilisation around at that time, including clean water, fresh fruit and a market place.

It managed to be in this wonderful position because of another change in the world's weather. Around 300BCE the weather turned warm and moist, the seas were calmer and the skies clearer; these were ideal weather conditions because they encouraged larger crop yields and easier trade routes on which to send crops and other booming produce.

In addition, this warming came at exactly the right time for the small Italian community known as the Romans. The Alpine mountains to their north suddenly became passable and they were able to take an army beyond its frontiers and as far as their resources and will power would take them; from Scotland to the Sahara. Wild fluctuations in weather during the period ensured that Rome's rise to glory did not follow a completely smooth path: in the summer of 54 BCE Julius Caesar had a long wait with his army while a persistent north westerly wind delayed his invasion force from crossing the Channel to England. The same problem faced William of Normandy in 1066 CE until a more helpful wind direction arrived in mid-autumn.

At the same time, the ancient Chinese also found that their known world had opened up to them too and that their empire could flourish and dominate in a similar way to the Romans. The two great superpowers began to trade thanks to a calming in the Himalayas. This opened up trade routes, such as the old Silk Road, and the sea routes from the Far East. It seemed that every major trade route passed through one special crossroads: Petra. In short, Petra was extremely successful because it was in the right place at the right time; thanks to the ever-changing climate.

But yet again, the same pattern of climatic change had a decisive influence on the fate of this civilisation; between around 300 and 800 CE the global temperature fell once again and the whole region surrounding Petra became horribly arid and dry. This drop also had a negative effect on the rest of the Roman Empire; drought developed on such a scale along the Great Silk Road that it actually stopped traffic along this route.

Way up in Central Asia lived the Huns – a nation-less gather of mostly sheepherders, so often described in history books as 'barbarians'. The dip in temperature meant freezing drought and drying of pastureland, and this forced the Huns to abandon their home and move southwest into the territories of their neighbours. A chain reaction was set off and soon dozens of large tribes across Eastern and Northern Europe were attacking the borders of the Roman Empire. Eventually the thinly spread Roman resources were undermined and the empire crumbled slowly into nothingness. So not only had the flourishing wealth and idealism of Petra been forced out of existence by the climate, but now the greatest and most powerful empire in history – until the rise of the British – was destroyed over a matter of decades due to the pressures of weather on human society. All does not bode well for the theory that power and security go hand in hand.

Greenland

It was perhaps the annual migrations of wild geese to and from Iceland and the Arctic that convinced a group of Irish monks – disgruntled with the general decline of the times and the barbarian migrations in Europe – that a land of peaceful retreat lay in the North. It was the Faeroe Islands on which they established themselves, as early as 700-725 CE, predating that of the Northmen[1] in 800, who forced the Irish to return home.

The first expedition of Northmen to Iceland came in 860, and the settlements seem to date back to the decade following. The period of great Norse voyages into the North Atlantic, eventually reaching as far as Newfoundland and Labrador, probably owes a lot to a lack of severe storms and a retreat of the sea ice brought about by the warm interlude at this time.[ii]

By 982 CE, a Northman called Erik the Red was banished from Iceland for a series of murders. He took a ship into the West, towards the little known land mass known to be around there. So far, Greenland was one of the least visited places on the planet because of the impossible terrain (heavy and unpredictable pack ice) that isolated it from Northern Europe. However, the

[1] Also known as Vikings, although that word describes an activity rather than a people or tribe – to go 'viking' is to go raiding, and the people commonly referred to as such were more like invaders and settlers than purely raiders. It is only the historians of the last few hundred years who have named the Northmen 'Vikings'.

temperature had jumped again since the fall of the Romans and the pack ice surrounding Greenland had melted, enabling Erik the Red to stumble upon a beautiful and lush land of grass and crystal clear water – hence the reason he named it "Green Land". He spent about three years exploring the land.

Once his banishment had expired Erik returned to Iceland and began gathering people from his homeland of Norway to come with him and help populate this paradise he had found. So, in 986 CE, 14 ships and 500 settlers landed in Greenland and set up two colonies on the West coast. There was plenty of pasture and grass for their animals and lots of fun and games for the newly independent people. However, one thing they didn't have was wood; if you ever make a trip to Greenland one of the first things you'd notice would be the astonishing lack of trees. To solve this problem, Eric's son set off to Baffin land, Labrador and maybe as far as New England, with a group of 35 men and returned with plenty of timber. These trips became regular events and the population of Greenland was dependent on them; some people even began building houses in 'Vinland' – or present day America – long before Columbus.

Of course, all would not remain so finely balanced for long, because around 1300 the temperature fell again, and by 1351 the sea ice was so thick that the sea route to Iceland was becoming blocked. Trade with Europe and Britain petered out as voyages were becoming dangerous and long – ships were often blown off course and never made it to port. Pope Alexander VI became greatly concerned about the situation in 1492, writing:

... the church of Garda is situated at the ends of the Earth in Greenland... and the people dwelling there are accustomed to living on dried fish and milk for lack of bread, wine and oil... shipping to that country is very infrequent because of the extensive freezing of the waters – no ship having put into shore, it is believed, for eighty years – or, if voyages happened to be made, it could have been, it is thought, only in the month of August...[iii]

Soon only one settlement remained on Greenland, and then later there was nothing left at all. Centuries later, a few of the original Greenland inhabitants were unearthed and a second apparent problem was discovered; not only had the Greenlanders frozen to death but they had also starved. Being the Europeans that they were, the Greenlanders were only used to eating meat and dressing in European clothes, and so when the temperature dropped they failed to imitate the Eskimos and wear sealskins and eat mostly fish. So because of the Greenlander's reluctance to adapt to the change, they starved to death; hence, Columbus was allowed, several centuries later, to 'discover' America, rather than just stumble across another Danish speaking colony.

The Norse colonies on Iceland suffered a severe decline, which may have begun around 1200 and lasted for a further six centuries. Tax records indicate that the population fell from around 77,500 in 1095, to about 50,000 in 1703.[iv]

From the 'Little Ice Age' to Industrialization

So far we've looked at how climate changes have single-handedly created and destroyed some of history's most successful human civilisations, but we have yet to fully prove the point that climatic changes can affect our modern world – after all, aren't our modern cities and societies much more stable than at any time in history?

Swings in global temperature did not stop after the fall of Rome and the advent of the Dark Ages. The next alteration of any significance began around 1650 to 1850 in North of Europe. Certainly, there may be strong ground to suggest that the bubonic plague – or 'Black Death' – which arrived in Europe around 1349, originated in Central Asia, during or after exceptional rainfall and flooding in 1332. This flooding was significant in itself: allegedly killing seven million people, making it one of the worst weather disasters in recorded history. The human settlements destroyed in the flooding also suffered the loss of their sewage systems; this and the loss of wildlife habitats, including those of rats, perhaps could have spawned the bubonic plague, which would go on to devastate Central Asia and Europe.

The middle of the sixteenth century saw a sharp fall in the temperatures of Europe, England and even California, suggesting a global significance. For the next hundred and fifty years the coldest period of temperatures occurred since the last major ice age ended approximately ten thousand years ago. Evidence of this cooling is distributed all over the world. This may be regarded as the peak of the Little Ice Age, although the entire period from 1420 to 1850 could have witnessed the inset and the outset of the Little Ice Age.

Records from sources in mainland Europe show an increasing tendency for wetter and cooler weather throughout the region; the summers of the 1570s and 1810s, and the springs of the 1690s, stand out as being particularly cold. A weather diary kept in Zurich between 1546 and 1576 shows that the relative frequency of snow, during the precipitation and snowfall of winter, was 44 per cent until 1563, and then 63 per cent from that point onwards.[v] Wind directions also changed, and there was an increasing extent of the winter sea ice around the northern coast of Scandinavia. Undoubtedly, it was these periods of cooler summers and shorter growing seasons that reduced grain production across Europe and Britain, and resulted in increasing grain prices between around 1510 and 1630. The cold, wet trend of the years 1570 to 1600 and 1690 to 1740 inclusive created big advances in the Alpine glaciers.

Northern Europe hosted some of the worst storms ever recorded during this epoch, becoming increasingly troublesome between about 1560 and 1720. These storms not only wreaked havoc directly, through gales and rainfall, but also brought about vast flooding in addition; parts of coastal Denmark, the Netherlands and Germany were permanently lost to the seawaters in the storm of October 21st 1634. The great storm which passed over southern Britain on the seventh and eighth of December 1703 wrecked houses and settlements all along the east coast of England, and demolished the Eddystone lighthouse near present-day Plymouth – the damage to London town alone was estimated at the time at £2 million, a huge amount in those days. Ships were wrecked along the coast or at sea, blown far up river, or lifted well beyond the usual reaches of the tide and grounded. Throughout the series of storms, thousands of lives were lost, many of which were the result of drowning or the spread of illness through the inland waters.

Cod helps to indicate the state of fishing throughout the North Sea, Baltic and off the coast of Scandinavia all through Little Ice Age period. The fish thrives best in waters between 4°C and 7°C and avoids waters colder than 2°C with great trepidation because it makes their kidneys fail and they die. The fact that this fish cannot be found in waters colder than 2°C help us to gather a rough idea of the water temperatures in almost any time period. For example, when cod migrated away from the Faeroe Islands about 1615, and in the same year that Iceland was surrounded by ice – 1695 – the cod became scarce around Shetland, and absent from the coast of Norway. This indicates that seawater temperatures were significantly lower (perhaps around 2 or 3°C) than fisheries were traditionally accustomed to, and much colder than today (around 5°C colder).

The shift in climate that would thaw the Little Ice Age, around 1650 to the early 1700s, sparked the boom of technology and the mass use of natural resources for the profit of man, in what later became known as the Industrial Revolution. Today, in the 2000s, the world is radically different from what it was like before this 'revolution' had taken place. We live today with consequences of that hundred and fifty or so years of technological and social upheaval, between 1700 and 1850; some of these consequences are very positive, some are very negative. Increasingly, in today's world we are ever more concerned about the climate change caused by human actions, and we are aware that this change may bring our downfall. Therefore, it is almost ironic that the very system that allowed us the opportunity to become a successful and developed civilisation is the same system that we ourselves are using to destroy our civilisation.

The warming trend that saw off the Little Ice Age arrived with as little sturdiness as the cooling leading into the period had set in. Throughout the eighteenth and nineteenth centuries the decline of the Little Ice Age was marred by hasty and erratic comebacks of cold weather. The winter of 1708 was historically brutal: seeing the Baltic ice over, allowing people to walk across its entire breadth; seeing the cultivation of vineyards in France abandoned because of the severity of the frost, eventually killing great numbers of trees across the country; and seeing ice develop all along the length of the Flanders coastline. In 1716, the river Thames was frozen again – like at the height of the Little Ice Age – and the ice was so thick in places that it stayed well into the new year; there was so much activity and games on the ice that London theatres were virtually deserted.

Despite a notable run of terrible harvesting years during the Little Ice Age era, there were also some particularly good ones – most notably during the 1490s, 1537-48, 1685-90, the period between 1700 and 1707, and a general proportional improvement in harvests from 1717 onwards. Indeed, when wetter springs and summers did persist, and led to poor wheat turnover throughout Europe and the British Isles, the discovery of the potato in South America, and its subsequent adoption by European countries, helped to fend off the famine. After the famine of 1772, the Hungarian government ordered the growing of potatoes as a preventative measure against future wheat harvest failure; so too in Russia during the 1760s and 1830s.

By 1750, England was enjoying particularly glorious weather; the days were brighter, the nights were calmer and the flora was blooming. It also happened to be perfect conditions for growing corn, and people had been noticing this for decades. In fact, the nation had been journeying into pleasantness for about 1000 years – due to a gradual temperature increase of 4°C - and when they weren't fighting battles and suffering plagues, the population was concentrating on ever increasing crop output, even in times before the cold spell of the Little Ice Age. By 1750, the crop yields were four times what anyone was traditionally accustomed to – suddenly people had more money. As the summers became hotter and moister, and the winters became milder, a handful of people realised they could experiment with technology to help make their lives a little easier.

A man named Abraham Darby (1678-1717) witnessed the families around him become healthier and more fruitful. The boosted crop yields had been followed by a booming population and a man like Darby knew how he could turn this unprecedented circumstance to his advantage. Darby made cast-iron cooking pots to help feed the blossoming masses but the traditional method for doing so seemed, to Darby, to be far too slow and inefficient. He needed fires and plenty of them, but the fuelling process he employed was nothing more than introducing wood and a furnace. Wood, in England, wasn't - and still isn't - particularly abundant, and Darby was stuck without an effective and cheap way of melting his iron. Ingeniously, he invented a way of purifying coal (which was very abundant in England and especially around the site of his new factory in 'Coalbrookdale') to make 'coke'.

Coke took off and so did the economy. The invention of the steam engine is greatly indebted to the existence of coke as a fuel source and, of course, without the steam engine there would be no manufacturing process and no factories. Factory jobs attracted more and more peasants off the land and into towns, and it wasn't long before the whole country was experiencing its first real economic boom – *anyone's* first economic boom.

The story of the Industrial Revolution is not so simple as to credit it to one man. In fact, it would be easier to credit the Industrial Revolution to the fuel that fired the furnaces than the men who put the fuel in. All of humanity's recent history has revolved around fossil fuels. The story of fossil fuels starts much earlier than you would think, well before the advent of the Industrial Age. It was in the England of Edward I (reigned 1272 to 1307 - the man whose attempts to invade Scotland were stopped so famously by the likes of William Wallace and Robert the Bruce) when coal began to be burned for fuel. If you've ever burnt coal in your home it will come as no surprise that it took so long to catch on – it produced thick, black smoke, gave off an acrid smell, and was traditionally much harder to come by than a good old bit of wood. Coal never caught on in vast quantities because wood was superior in so many ways: cleaner, more abundant, had a more pleasant smell when burnt etc. In fact, in 1306 King Edward himself banned the burning of coal throughout his kingdom because he couldn't stand the smell.[vi]

However, by this time the great forests of England were fast becoming a rare sight. For centuries people had been taking the trees and forests around them for granted, building houses and forts, constructing walls and lighting fires. By the time of Edward I the price of wood was rising fast, and alternatives had to be found unless the King wanted to watch thousands of his poorest subjects die every winter due to lack of warmth. England soon relented and became the first country to burn coal on a large scale.

It is only fairly recently when people began to understand what coal was and how it was formed. Six or seven centuries ago nobody had a clue, and many associated it with disease, or attributed its existence to the devil himself. What could be more devilish than a dirty, black rock that smelled of brimstone when burnt and reminded everyone of the Hell that possibly awaited them?

Nevertheless, coal would gradually gain momentum as an energy source in the British Isles and, by the early eighteenth century, over a thousand tonnes of the stuff was being used every day in London. There was also a rising demand for a source of fuel to power the factories springing up all over the country. The traditional method of energy generation was the water wheel, meaning mills had to be placed as close to rivers as was humanly possible. As the products of such mills began zipping out of the doors at a faster and faster rate, it became clear that the power of water was only going to take production so far. If another principle source of energy wasn't found to supplant water and wood as England's primary fuel sources, then a great energy crisis would befall the nation.

Coal, however, wasn't in the best position to step up and take the reins. By the 1700s English coal mines (and other mines) were suffering from flooding, with water coming in from the rain above and the water tables below. The water had to go or else mines up and down the country would become nothing but useless holes in the ground. Luckily, there was a man with an idea clever enough to tackle this problem. Thomas Newcomen (1663 – 1729) was a simple ironmonger from the sleepy town of Dartmouth, but he managed to solve the water problem by devising a pump, for which he used a moving piston. To move the piston he burnt coal, which boiled water, produced steam, created a vacuum by condensing, and finally moved the piston back and forth. His idea took influence from an earlier invention, a forerunner to the steam engine, called the 'atmosphere engine' by its creator Denis Papin (1647 – c. 1712), who also invented the first pressure cooker. The first Newcomen engine was installed in a Staffordshire coal mine in 1712 and was initially used largely by the tin mines of Cornwall; by 1750 there were hundreds of them employed at mines all across the country, and more than 6 million tonnes of the black stuff were being dug up every year.

Other inventions and innovations pushed the Industrial Revolution onward. The flying shuttle was invented in the early 1720s by another Englishman named John Kay (1733 – 1764) and succeeded in mechanising the cotton weaving process. Cast steel was first produced by Benjamin Huntsman (1704-1776) in the early 1740s, also in England; bar iron was purified in the production of steel by intense heat treatment. These innovations, and many more like them, demanded additional energy and produced other industries and businesses that in turn demanded *more* energy. Coke was proving a saviour for a world surging forward relentlessly.

The steam engine was transformed by James Watt, who first improved on it and enabled it for wider purposes, and who then, in 1784, made it mobile. The era of the steam train was born, and when trains were adapted to take human passengers from place to place, rather than just raw materials, the human race had taken its final big step into a new world. Because of its unmatched ability to be used for heating, cooking, pumping, industry, and now transport, coal became the lord of this new world, powering humankind full steam into the 1800s. Humanity was now on the move, in more ways than one.

England wasn't going alone into the new world - several other rich countries followed suit harnessing the fossil fuel technology, and by the mid

1800's it had infected the fledgling United States of America. One of the residents of the USA was a certain Thomas Edison – a man obsessed with invention and progress; he once said that he couldn't pick anything up without wanting to improve upon it. Edison had left work and set up his own industrial laboratory – the world's first – and began experimenting with anything new he could think of. The most significant invention that he and his team came up with was an artificial light source also known as a light bulb. Now it was possible to light houses and offices and factories and palaces without firelight, candle light or larger than life windows. In 1882 Edison opened the world's first power station for electrical lights in lower Manhattan, dawning a new era of power generation, and resurgence in economic industrial advance.[vii] In essence, Edison helped develop electrical systems everywhere by creating a demand for something out of nowhere. Edison had made electricity useful.

Though electricity was to become the jewel in the crown of the Industrial Age, its mass production was only available thanks to industrial processes. Before you can feed an electricity grid you need to create it in the first place, and to do that you need miles and miles of railways and plenty of the substance many called "black gold"… otherwise known as coal.

---

[i] Lamb, H. H. (1995). *Climate History and the Modern World, 2nd Edition.* Routledge, New York. Pg 154.

[ii] Ibid. Pg 172-177.

[iii] Stefansson, V. (1943). *Greenland.* George Harrap, London, Toronto, Bombay and Sydney. Pg 240.

[iv] Lamb, H. H. (1995). *Climate History and the Modern World, 2nd Edition.* Routledge, New York. Pg 189.

[v] Ibid. Pg 212-213

[vi] Flannery, T. (2006). *The Weather Makers.* Penguin Books Ltd, London. Pg 73.

[vii] Freese, B. (2003). *Coal: A Human History.* Perseus Publishing, Cambridge, Massachusetts

PART 3 - the Age of the Human

By 1910, two inventions were revolutionising the Revolution:

1. The manufacturing of spare parts created new markets;

2. Mass production - America was the first to witness the new phenomenon of 'the production line', where thousands upon thousands of goods could be constructed within one working day.

In addition, the development of propaganda during the two World Wars served as a lesson to industrialists who wanted to create mass markets for their everyday products. Edward Bernays - the nephew of Sigmund Freud - using his uncle's theories, became the first to attach consumer products to emotion and happiness. Suddenly the average person was buying products because they wanted them, *desired* them, and not simply because they needed them. Coupled with the manufacturing of spare parts and the birth of the production line and very soon, 'consumerism' was born.

Developments such as these at the beginning of the twentieth century have led us out of the Holocene and into what scientists call the 'Anthropocene' or the Age of the Human. The vital systems of the planet are being disrupted by our species – the carbon cycle, the nitrogen cycle, the ozone layer, ocean circulation and the ice caps. We are now a physical climate force of our own. Exactly *how* is the next question.

Chapter 6 – Humans and Climate

Consider this: the Earth is 4,500 million years old – the first human-like primates to walk on two legs and use tools made from stone only began to emerge 4 million years ago. In other words, our 4 million year existence as a bipedal species (estimated) is only worth one ten thousandth (1/10000) of Earth's existence as a planet.

Furthermore, if we want to get more specific, we find that our species – Homo sapiens – has only been around for roughly 100,000 years or so. So we must wait until 99.9978 per cent of Earth's history has passed before we begin to see the first people arrive. These numbers get even larger when we consider that it is only within the last 10,000 years that human civilisations have begun to develop, and that it is only within a couple of hundred years that human society begins to resemble what it is today.

Though our species has managed to survive the many pitfalls it has faced since its early years, we still face a dangerous future. An asteroid similar to the one that struck 65Ma may hit us, or we may fall victim to a super-volcanic explosion that blackens the sky and poisons the air for thousands of years. Either way, our climate will eventually change – naturally – away from this momentary bliss of the Holocene. We may even evolve ourselves into separate species given a long enough time (and by all the laws of evolution this seems inevitable). And even if we survive all that, in about 6,000 million years the Sun's energy will be spent and it will swell into a red giant and engulf the inner planets – including Earth. By all long-long-term forecasts, the human race seems doomed.

Philosophers have been speculating about the meaning of life for thousands of years but it is only present-day scholars who have the ability to take into consideration man's limited lifespan on this planet. When we take the fact that our species will one-day end (by catastrophe or evolution), our philosophical window opens wider. Where are we going and how can we get there? How can we navigate this minefield of disasters and successfully keep our societies and our species intact?

Surely one of the most obvious steps towards longevity is looking after our home. If we properly care and respect the planet we are on, the likelihood is that we will do better as a species than if we just exploit and pollute it. Are we going to be parasitic consumers or find a balance with nature?

We are already facing several problems, especially the issue of human-induced climate change. And here we come across an interesting point: how on Earth can we expect climate change to be a human-caused issue? Surely the climate is too powerful and complex to be changed by a single species - 'human-induced climate change' must be something of an oxymoron. To understand this we must first look at the history of human-climate interactions.

Increases in CO_2

In Part 2 we saw how the Earth's ever-changing climate has influenced the course of human history. But at some point in time, probably around 200-250 years ago, the tables began to turn. Gradually, human activities began to affect the climate as well as just the other way round.

Early humans affected their world by cutting wood for shelter, digging holes, burning vegetation, farming, hunting and so on. Similarly, man-made damming of rivers creates artificial lakes today that provide additional water vapour to the atmosphere, which falls as rain in that local area. Asphalt surfaces soak up more sunlight during the day and release it back into the atmosphere during the night as it cools. In other words, human existence having an effect on the environment is unavoidable, and almost natural to a certain extent. But whereas asphalt surfaces and artificial lakes are somewhat necessary for providing extremely large populations with energy, communication, transport links and fresh water, not everything in this human world that damages the Earth is so localised and justifiable.

Carbon dioxide is probably the most important gas in the atmosphere, used to bring about climate changes and climate stability, over both short and long time periods. Ice core evidence of historic CO_2 concentration in the atmosphere presents us with evidence of human pressure on this careful and fragile system. Each bubble of air trapped in an ice core is a historic and priceless time capsule, holding information about what the atmosphere was like at the time the bubble formed. As each CO_2 concentration point is graphed, an obvious and shocking pattern is revealed.

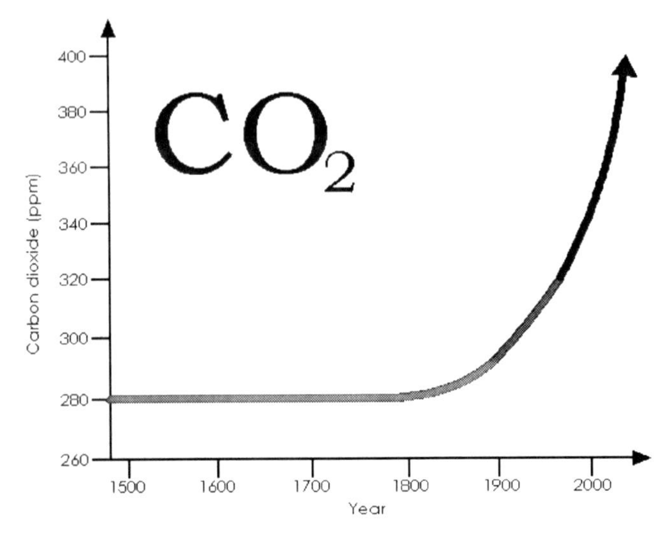

Figure 6.1 (above) – The recent story of atmospheric carbon dioxide levels is not one of subtlety, even when smoothed into a trend curve.

Since the last glacial maximum, the level of CO_2 in the atmosphere has remained at an almost constant value of 280 parts per million (weighing roughly 590 gigatonnes – thousand million tonnes) for several thousand years; this is known as *the pre-industrial CO_2 level.* Then, at around 1800 CE, the amount of CO_2 in the atmosphere began a steady increase. By 1850 the concentration was more than 285ppm and by 1900 it reached almost 300ppm.

By the turn of the millennium, in 2001, atmospheric concentrations were greater than 366ppm. The difference of a total of 86ppm from the pre-industrial CO_2 level is an *increase of 30 per cent*. The rate of acceleration is on a constant increase and subsequently 70 per cent of that rise has occurred in the last fifty years.[i] Climatologists call this concentration increase as the *anthropogenic CO_2 increase*, because it results from human activities putting extra carbon into the atmosphere. Today figures of atmospheric concentration of CO_2 stand at around 380ppm (equal to 785 gigatonnes). To say that nature never saw fit to push CO_2 levels beyond 280ppm for thousands of years before the Industrial Revolution, it is something that humans have managed to do quickly, easily and with no apparent guilt.

Two human activities have been mainly responsible for this extra carbon input: land clearance and the burning of fossil fuels - the first of these began during the 1700s and on into the 1800s. Trees have been cleared to make way for humans well before history began to be recorded. Originally, most trees were cut to make various tools, building materials, firewood and so on. But in the early eighteenth century in Britain there was an increase in demand for firewood and charcoal. By the time the industrial revolution had kicked off an era of unprecedented economic boom, the amount of carbon released into the atmosphere by human activity was over a gigatonne every year, from virtually zero.

Then the second carbon producing human activity emerged around 1800: burning fossil fuels. Fossil fuels are exactly what they sound like in that they are dead, fossilised organisms, burnt for the purpose of powering machines. These organisms were, essentially, plants that drew carbon from the atmosphere in photosynthesis and used it to form their stems, leaves, roots and other plant bits and pieces. When we burn fossil fuels we are liberating carbon that has been kept deep underground for millions upon millions of years, all of it once used in the building blocks of plant-life.

There are three types of fossil fuels, all containing carbon: coal, oil and gas. All have varying abilities to add carbon dioxide to the atmosphere. The blackest of black coal is almost pure carbon; one tonne produces more than 3.5 tonnes of CO_2 when it is burnt. Oil-based fuels are not so carbon-rich and produce less CO_2 per unit burnt; gas contains even less, meaning it is the cleanest of all the fossil fuels.

Coal is as common as muck. You can find it on all continents, and where it lies there is usually a lot of it. It forms when trees die, fall over and sink into a swamp, and where their rotting is impeded by a lack of oxygen. Vegetation builds and builds as it dies and falls to the ground, creating a thick soup of saturated plant matter. When nearby rivers wash sand and silt into the swamp the vegetation becomes compressed, pushing out the moisture. As the swamp is covered in earth and buried deeper, both pressure and time combine to make the organic matter more solidified and dense. Left long enough, the organic matter will convert to the black substance we currently see in our head as coal, and eventually becomes a solid, shiny jewel of jet black. As its name suggests, the Carboniferous period, 360Ma to 290Ma, saw the creation of most of the major coal deposits deep underground, during a time when the world was largely covered in tropical and semi-tropical forest and plenty of swamps.

Oil, on the other hand, is far less abundant than coal because it forms from life in oceans and river estuaries. Oil is largely composed of dead phytoplankton, which we already know is a single-celled, ocean-based plant that can have a huge influence on global climate. Though oceans and seas seem abundant on Earth, oil remains a rare commodity because it is liquid, and with a dynamic and ever-moving tectonic crust, oil can be lost in the lithosphere if it isn't trapped in a specific way by surrounding rocks. Secondly, oil has a very precise recipe for formation. The organic matter has to be 'cooked' at around 120°C for millions of years – any more than 135°C and all you get is natural gas.

Since its transformation from a dirty, black substance to a wonderful source of thermal energy by such figures as Newcomen, Watt, Darby and Edison, coal has never stopped being demanded by our societies. Despite its associations with long-gone industrial times, polluted cityscapes and factory chimneys of the 1800s, more coal is burned today than at any other time in history. And the future trend appears to be continuing, with 249 coal-fired power stations being built around the world between 1999 and 2009, and about 1200 more by 2030. A major contributor to the extended life span of coal is China, which is looking to take advantage of huge coal reserves running right across vast areas of its land mass. China will build more than 500 coal power stations before 2030 as it seeks to provide electricity for its booming economy as cheaply as it possibly can.

Applying a perspective to these figures can strike the fear of God into even the most apathetic conservative. First, take into account that the total number of coal power stations yet to be built between now and 2030 will produce around 700,000 megawatts of energy. That's 2,500 thousand million tonnes of carbon shoved into the atmosphere. Then, take into account that the average lifespan of a coal power station is 50 years, and that it will take centuries for the CO_2 they produce to stop warming the atmosphere – long after they are dismantled – and we have a really big issue on our hands.

Even if you ignore the huge carbon emissions, coal isn't even efficient or clean to burn. It comprises of between 70 and 90 per cent carbon, one and two per cent nitrogen and 0.5 and five per cent sulphur, amongst hydrogen and oxygen as well. Sulphur is an extremely toxic pollutant when converted into sulphur dioxide (which it is in a coal power station), being harmful to humans as well as other ecosystems. Though the efficiency of conversion of chemical energy in fuel to heat is more than 90 per cent, the overall conversion efficiency of a coal power plant is around 35 per cent. This is because heat is poorly converted into mechanical energy, a process that cannot be made much more efficient, even by the sharpest of technical minds. The chemical composition of stack emissions from a typical coal power station is: 80 per cent oxygen depleted air, 12 per cent carbon dioxide, 4.5 per cent water vapour, and around 1700ppm sulphur dioxide. Additionally there are smaller emissions of what are known as nitrous oxides (around 650ppm). With the greenhouse gas carbon dioxide, the sulphur dioxide and nitrous oxides go on to create acid rain, an environmental gremlin for humans and animals alike.

Coal powered the 1800s with great gusto, but by 1900 it reluctantly handed the baton to oil. The twentieth century was without doubt the Century of Oil. Discoveries in North America and the Near East during the early part of the century paved the way for the dawn of the new energy era. The Near East is key to the Oil Era; until it was tapped, one in seven barrels of the world's oil lay under Arabia. Because of this, the entire region remains the focal point of all global political agenda, with colonisation, imperialism and countless wars taking place there since the turn of the twentieth century. The Rich North, led first by Britain and now by the United States, has fought for decades to control the region, with certain successes and failures along the way. The fingerprint is unmistakable. Rich Northern interests in this region only developed in the twentieth century, becoming more and more strategic and political as the century went on. Before oil, the region was seen as merely a strategic significance by the world's most powerful countries, with little to offer in terms of resources.

Until the 1960s, the world's oil companies were discovering more oil every year. Since then, less and less has been discovered, mostly due to the fact that we'd already found most of it. Many underground oil sources are not even worth tapping as they are quite small, meaning they cost more to drill and extract than the value of the oil held there would generate in revenue. By 1995, over 24 thousand million barrels of oil were being used every year on this planet for one purpose or another. Today, the cost of oil can be enormous compared with what it has been historically, and most experts agree that it'll only get more expensive as time goes on and reserves run out. By as early as 2040 oil may become so rare that it is no longer economically viable (meaning it is not worth bothering with), forcing us once again to turn to a new primary source of energy.

By 2025 that new source could be natural gas. In 2002, humans produced 21,000 million tonnes of CO_2, which was released directly into the atmosphere. Of this total, coal contributed 41 per cent, oil 39 per cent and natural gas 20 per cent. Though gas only took over coal in importance to human energy production around the new millennium, it could be only a couple of decades before it supersedes oil.

And in case there was any doubt to as exactly how much fossil fuels still matter to us, in 2003, despite fierce international condemnation, reluctance by the United Nations to legalise it, and no physical proof of a threat, the US and UK led a small coalition of nations into Iraq to topple the leadership. Just days after the invasion was complete, the oil was once again being pumped, with special contracts being passed to American corporations seeking to do business in the oil fields. The whole ordeal proved only what many had known for a long time – oil is as vital as ever to the movement of Rich economies, and until the thirst for oil is quenched, the Near East will remain the very centre of the political world.

Before humans arrived in the British Isles, about five thousand years ago, the land was vastly covered in woodland; where moors now lie, once there were trees. And sure enough, the UK has a relatively large reserve of coal locked below the ground, evidence that once great forests stood above. More proof is found in the tropical rainforests, where great and expansive jungles cover much of the land. Below the rainforests are great reservoirs of oil and gas, the decay of millions of years of dense vegetation sitting above.

Electrical energy cannot be created without energy being put into the process to begin with. Today we can harness the energy in the wind, the sea and the sun to create electrical energy, but in 1800 the only option was the good old-fashioned idea of burning things. Coal was initially the only fossil fuel used but eventually oil and then natural gas were also burnt. By the early 1910s the majority of the carbon released into the atmosphere was from burning such fuels, and it has remained that way ever since. Figure 6.2 below illustrates these two factors against time and shows the dramatic increase of total annual carbon emissions from virtually nil in 1700 to one gigatonne by 1800, over two gigatonnes by 1900, and then to about eight gigatonnes by 2000. The vast majority of this is due to burning fossil fuels and releasing carbon into the atmosphere, which is supposed to be locked deep in the lithosphere. The small amount of carbon produced from land clearance is mostly due to the removal and burning of tropical rainforest.

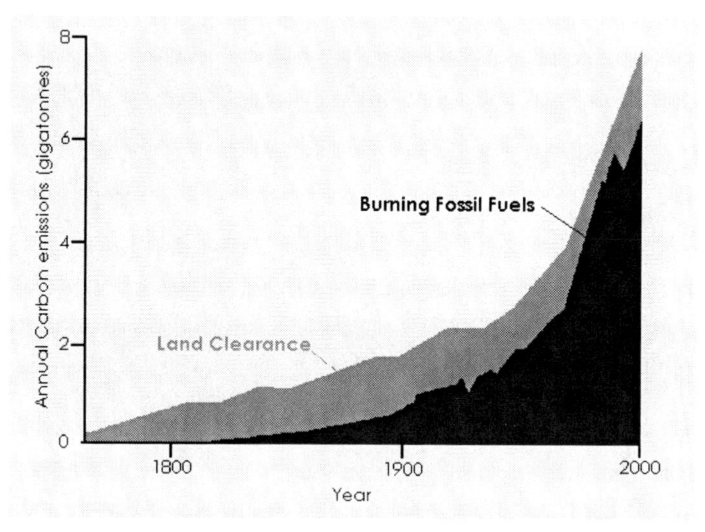

Figure 6.2 – Human sources of carbon dioxide emissions since 1700: land clearance and burning fossil fuels. Both have surged in the last few decades but it is the fossil fuel use that is the major problem.

Atmospheric CO_2 levels have increased by 30 per cent since 1800, merely two hundred years or so. So what has happened to this additional carbon? The answer to this is dependent on things known as 'carbon sinks' – components of the planet's system that absorb atmospheric CO_2. As we already know, carbon cycles between the atmosphere, the lithosphere, the ocean and the biosphere. Thankfully, not all the CO_2 is left in the atmosphere – rather this figure is about 55 per cent. Interestingly, there is a blinding clue in the case of 'who-dunnit' within the atmosphere itself; recordings of atmospheric CO_2 have shown that more exists over the Northern Hemisphere than over the

Southern, matching human population distribution. If humans did not have such a vast influence over CO_2 in the atmosphere then its concentration would be in more natural locations, such as over forests and over tectonic boundaries, where the gas is frequently emitted in large quantities.

Calculations show that about 25 per cent of the excess carbon dioxide is absorbed from the atmosphere by the ocean. Oceans have absorbed 48 per cent of all carbon emitted by humans over the last two centuries – by far the largest sink on Earth. However, because the ocean is so deep, and its waters mix rather slowly, most of the oceanic CO_2 lies within the upper surface waters - far less the deeper you get. Also, carbon dioxide gas dissolves more easily in cold waters than warm. As a result, CO_2 is absorbed more in waters further from the warm equator, such as the North and South Atlantic, and North and South Pacific. In warmer waters, CO_2 is actually released back into the atmosphere – the central Pacific being the biggest source.

Furthermore, it seems the future of the great ocean carbon sink is far from assured. Scientists now predict that climate change will degrade its ability to absorb carbon over the next hundred years or so, as the oceans warm. This is no more important than in the North Sea - the little patch of water above the British Isles and to the west of Scandinavia. Here scientists have discovered that carbon is being absorbed at a huge rate, because of the uniqueness of its position and its water currents – it is thought that the North Sea could have absorbed around 20 per cent of all carbon dioxide ever emitted by human beings. Warmer North Sea waters would seriously damage this system, dramatically reducing its lust for carbon.

If this wasn't enough, it seems the great ocean sink is already growing considerably weaker. In the 1980s oceans were zapping 1.8 gigatonnes of carbon from the atmosphere, but by the 1990s they were taking only 1.6 gigatonnes.[ii] Organisms in oceans that make carbonate shells (and thereafter transfer carbon to the sea floor upon death) become increasingly poor at this job the more waters are acidified, as they would be under climate change conditions. By 2100, the amount of carbon taken in by the oceans could be only 90 per cent of what it is today.

The remaining 20 per cent of CO_2 must be absorbed by the biosphere, with the lithosphere taking very little directly from the atmosphere. Reforestation projects in parts of the developed world have helped take in a sizeable volume of human carbon emissions. However, more interesting is the volume of carbon that is absorbed by vegetation which is not part of the reforestation. This is done through something known as CO_2 fertilization: the process of photosynthesis, employing carbon taken from the air to build physical components of vegetation (leaves, trunks etc). Experiments have shown that most vegetation grows faster, and more densely, in an atmosphere containing more CO_2 than less. It has also been discovered that the density of Amazonian vegetation had increased in recent decades, suggesting that the massive area of forestland was getting healthier, due to additional atmospheric CO_2. Vegetation was growing rapidly, and trees appeared to have longer, healthier and more numerous leaves and branches; generally speaking, the Amazon Rainforest – aside from areas where deforestation is occurring – appears to be gaining strength.

Superficially, therefore, it seems that additional carbon dioxide in the atmosphere actually makes dense vegetation zones more dense, encourages life, and improves the overall health of some plant life on Earth. However, as we will find out later on, if there is too much additional carbon dioxide, plants tend to feel a little overwhelmed and die – releasing whatever CO_2 they captured back into the atmosphere. Ultimately, this carbon sink becomes a carbon source, providing positive feedback to global warming; this is only the tip of the iceberg - no pun intended.

Increasing Methane

Carbon dioxide is not the only greenhouse gas. Equally, it is not the only greenhouse gas that has been produced by human beings over the last two centuries; there are handfuls more. Arguably the second most significant of those gases is Methane. Formed from a relatively strong bond between one carbon atom and four hydrogen atoms (CH_4), methane is an undeniably natural gas. In fact, methane is produced in any breakdown of vegetation in an area lacking oxygen: from bogs and swamps to the digestive systems of cattle and… yes, even humans. Carbon that has been lucky enough to become part of some leaves or shoots, may soon find itself journeying through the gut of some creature.

CH_4 emissions have risen dramatically since the beginning of the industrial revolution 250 years ago – today they are about 145 per cent above natural levels.[iii] Prior to this period, according to polar ice core measurements, CH_4 levels were more or less stable at around 700 ppb (parts per thousand million). Like the increase in CO_2 since 1800, the rise in methane slowly accelerated throughout the nineteenth century, followed by a booming acceleration throughout the twentieth. Recent measurements give today's methane concentration to be at around 1800 ppb: more than 250 per cent of what it was just two centuries ago.

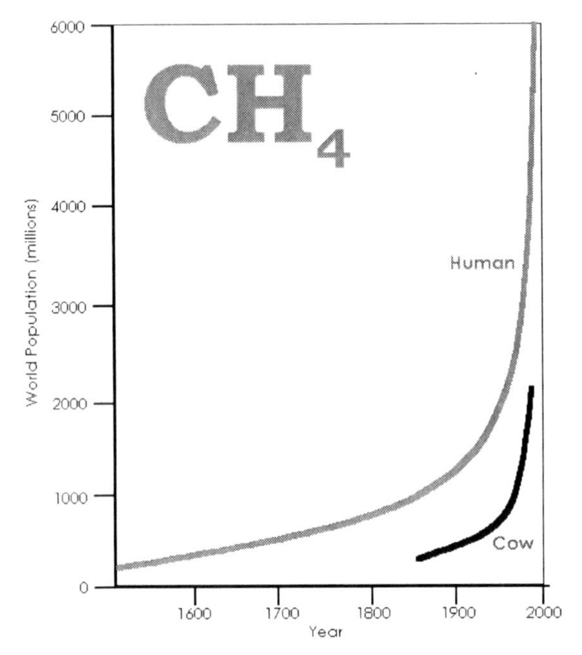

Figure 6.3 – The boom in human population over the last hundred years has been shadowed recently by a boom in cattle population. Both booms result in more methane emissions.

The reasons behind these increases are not as much to do with energy production and land clearance as those of CO_2 increases. It turns out that just existing is enough to push up methane levels anyway. Unfortunately, since the industrial revolution, the global human population has been increasing, and each new person brings new demand for food. In Asia, vast areas of tropical wetlands are being transformed into rice paddies, and this contributes to methane output. Additionally, growing dependence on meat as a source of nutrition has increased the population of cattle throughout the world, particularly at the demand of Western markets. Farm livestock contribute to methane emissions immensely, and whilst agriculture is fading in many first-world countries, the majority of developing countries are increasingly dependent on farming.[iv] The global population of methane producing cattle has mirrored that of humans over the last 50 years or so. This is a problem caused not so much through pollution or industry, but rather the failure and imbalance of our global economic system: not only does it force the poor into mass agricultural markets, but a symptom of our present system is a booming human population in poor nations.

Of course, cattle are not the only animals to emit methane but populations of other mammals have not increased with such a bang as those of cattle, and over such a short period of time. Since 1850, human beings have sought to maximise and mass-produce the farming industry, resulting in higher and unnatural populations of livestock. This is the reason that livestock produce 60 gigatonnes of methane every year - and why we can say that this is purely the result of anthropogenic activity.

Chlorofluorocarbons and Ozone

The story behind one of the world's most infamous gases brings us to our first real study of something called Ozone. Ozone is a naturally-occurring gas and usually forms high up in the stratosphere. When ultra-violet radiation, coming from the Sun, hits oxygen molecules in the atmosphere (O_2) it can sometimes cause them to split into two single atoms ($O + O$). These free-wheeling atoms may then cling to other oxygen molecules to form a set of triplets (O_3), and this is called ozone. Ozone will not build up in the atmosphere to too large an extent, however, because similar mechanisms causing ozone to form also cause it to break up, back into oxygen again.

However, visible solar radiation also destroys ozone, and because this is more abundant than ultra-violet radiation, ozone is destroyed faster than it is produced, making it very short-lived. The process of ozone breakdown can be speeded up by the addition of other gases into the equation; including chlorine (Cl), fluorine (Fl), and bromine (Br), all occurring naturally on Earth in ocean salts. Chlorine is particularly good at destroying ozone, stealing one of the oxygen particles from the ozone molecule to produce Chlorine monoxide and oxygen, before breaking down again into just chlorine and oxygen. At this point the cycle starts all over again with the independent chlorine atoms re-attaching themselves to the ozone molecules; and it is as simple as that.

But as much as it is simple, this process is also essential to the well-being of all life-forms down of the surface. The Ozone Layer famously shields Earth's surface from harmful UV-B radiation arriving from the Sun, that can cause cell mutations in living things, including humans - causing such problems as skin-cancer.

Ozone is also beneficial to humans in another way, lower down in the troposphere, helping to remove sulphur dioxide (SO_2) and carbon monoxide (CO) from the air – both of which are damaging to human and animal respiratory systems. However, ozone can cause damage to the lungs and eyes if its surface concentration is anything above scarce.

James Lovelock, the man behind the Gaia Hypothesis, discovered that a group of gases called chlorofluorocarbons (CFCs) were steadily increasing in concentration in the atmosphere. CFCs do not occur naturally in the atmosphere, and are produced entirely by humans for use in such appliances as solvents, refrigerator and air-conditioner coolants, and foam insulation for buildings. Unfortunately, once released, these compounds can stay in the atmosphere for a whole century, during which time they manage to reach the upper atmosphere and, more importantly, the stratosphere.

After a period in which CFCs were considered completely harmless, a couple of atmospheric chemists named Mario Molina and Sherwood Roland hypothesised a connection between rising CFC concentrations and destruction of natural ozone. In the 1980s measurements of ozone over Antarctica confirmed that the pair was correct; there was indeed a connection between rising CFCs and falling ozone levels. The tests showed an unusually high volume of chlorine in the otherwise 'clean', fresh Antarctic air – during springtime, the amount of ozone over this spot had decreased by over a half, in just two decades. This was the consequence of CFC build-up in the cold polar air during winter, suddenly being bombarded with solar radiation when spring would arrive, and rapidly destroying any ozone that happened to be around. Scientists would call this area of dilute ozone the 'Ozone Hole' and soon would begin years of great interest and attention amongst the international press.

But, by the time the innocuous reputation of CFCs was disputed, their production had rocketed: in 1954 it was 75,000 tonnes; in 1974 it was more than 750,000. Atmospheric concentrations are around five times as much as this today - and whilst estimates warn that atmospheric chlorine concentrations should not exceed 2ppb in order to prevent further damage to the ozone, today there may be upwards of 6ppb.

The major impact of increased ultraviolet B (UV-B) radiation is a weakening of the human immune system, reducing the body's ability to naturally fight off infectious and fungal diseases. Furthermore, in areas where infectious diseases are widespread, and malnutrition is rife, the efficiency of some vaccines may be reduced. In poor countries, the malnourished and the elderly will become more susceptible to cataracts with increased UV-B exposure. When living cells absorb UV-B radiation, some essential molecules become damaged, including DNA, which keeps the cell functioning correctly. Skin cancer manifests after cumulative exposure and hence why it is more common in the elderly. If global ozone levels are reaching their lowest point around the present, and begin to gradually recover from the near future onwards, then the delay process could see the subsequent peak in skin cancers around 2050.

UV-B radiation also affects other life-forms on earth, and – most importantly – sea plankton. When plankton is killed across wide areas, not only are marine food chains affected, but less carbon dioxide is absorbed. Elsewhere, too much exposure can lead to genetic mutations in some crops, and stunted growth in many plants. Plant DNA is damaged when plant foliage is damaged.

Of course, this global concern over falling ozone levels only began once people realised that less ozone leads to problems like those given above, but, nevertheless, humanity's biggest ever international environmental challenge was about to start. In 1985 the British Antarctic Survey first reported evidence of the ozone hole and sped up negotiations considerably at the Vienna Convention for the Protection of the Ozone Layer happening that year. In 1987, every major industrial nation on the planet met in Montreal, Canada, to sign an agreement aiming to cut and then ultimately eliminate the production of CFCs. The provisions of the Protocol were scientifically proven to be inadequate, however, and so a revised schedule for the complete phasing out of ozone-depleting substances was set in 1997 – global diplomacy is rarely so rapid! This swift action by the international community averted serious cases of harm caused by the ozone 'hole', and raised awareness about the dangers of apathy towards environmental issues on an international stage.

However, whilst most businesses that once used CFCs have now found alternative substances, CFC levels continue to rise, mostly due to those parts of the world that have been slow to make changes. China in particular still mass-produces cheap refrigerators that use CFCs, and this could work against international efforts. Additionally, the 100-year lifetime of CFCs in the atmosphere mean that it may be a long while yet before levels begin to fall, regardless of any production. The illicit trade in CFCs is estimated to be around 25,000 tonnes a year still.[v]

While many hail the slowing of CFC production as a mighty success, it is hard not to feel pessimistic about the whole ordeal. For a start, the actual first practical steps of tackling the CFC problem occurred only when alternatives to CFCs were being found to be economically viable by manufacturers. Governments hardly took the lead on the issue at all, and it is a little naïve to suggest that manufacturers would have gone along with cutting out CFCs were the alternatives expensive and not within arms reach. It remains to be seen whether the international political arena is capable of averting large-scale environmental catastrophes against the wishes of the market economy. The battle against CFCs is not a valid precursor to the battle against climate change for exactly this reason; averting climate change is not simply a case of swapping one chemical for another.

Sulphate Aerosols

During the nineteenth and early twentieth centuries, it was common to wander through a major city in an industrial country and see blackness everywhere you went. Industrial processes, such as smelting and burning of coal, produced this blackening effect and made the city air thick, hazy and unhealthy. Many industrial cities of the era, such as Manchester, in North West England, developed a reputation for being poor places to live and work. The reason is down to soot and ash from chimneys, but for our purpose we'll just call it all sulphur dioxide (SO_2) - a by-product of those industrial processes – that reacts with water vapour producing sulphate aerosols.

These aerosols are dense and heavy, and stay within the lower few kilometres of the atmosphere, never moving very far from their original source. It wasn't until the end of World War Two that factory smokestacks were made much taller. The emissions were thus dispersed much higher into the air, where wind would catch them and carry them away from the heavily populated areas; today, cities like Manchester are a sparkling contrast to their former days. The genius and simplicity of the solution to the health hazard is not without its own undoing, for while the quality of inner-city air was drastically improved, the particles were now dispersed over a much broader area.

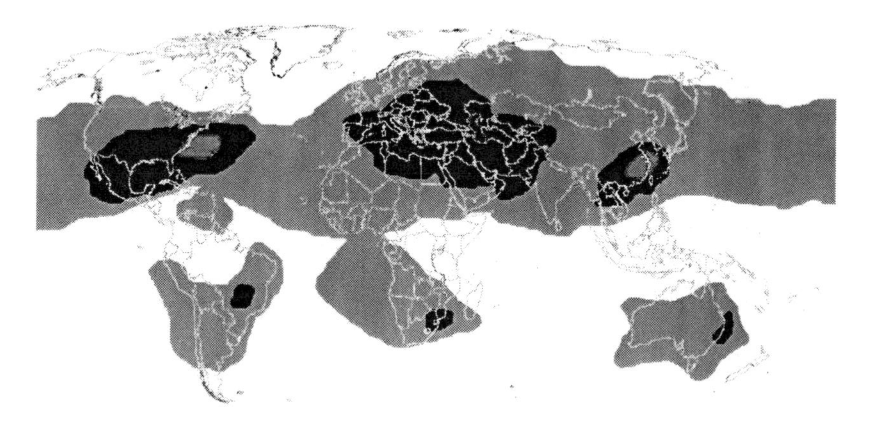

Figure 6.4 (above) – Sulphate aerosols exist in almost every part of the atmosphere, but in this image the highest concentrations are shown. This is also only a simplified view of non-volcanic world sulphates. The more intense concentrations are shown in blue, illustrating that most exist above the large industrial centres of the world, including South East Asia and the US.

Despite their relatively short lifespan (sulphate aerosols are removed by rain and, hence, remain only for a few days), these substances can still travel up to 600 kilometres from their source, thanks to just the wind. Consequently, large build-ups of sulphate aerosols exist to the east of Europe, the northeastern United States and over Eastern China.

Some debate exists as to whether sulphate aerosols actually encourage the global warming effect or work against it, cooling the atmosphere, but it is certain that they have a significant influence over climate change one way or another. The cooling speculation develops from two sources: first, the particles are big enough to reflect much solar radiation back into space and thus cool climates regionally; secondly, the particles attract atmospheric water vapour, creating droplets and eventually clouds. The bright surfaces of clouds reflect incoming solar radiation and keep local air temperatures cool. On the other hand, high atmospheric clouds trap outgoing radiation and should increase the temperatures on the planet's surface.

So which is right?

The net effect of sulphate aerosols appears to be cooling, and this, in theory, should offset global warming and work in our favour. However, a new phenomenon of 'Global Dimming' suggests that if we tackle sulphate pollution, in order to curb the rate of illness in human populations, then this would unleash the full drama of the global warming problem. In other words, our own pollution is saving us from experiencing the full effects of global warming. It is a very interesting conundrum, one that is definitely worth looking into in Part 6.

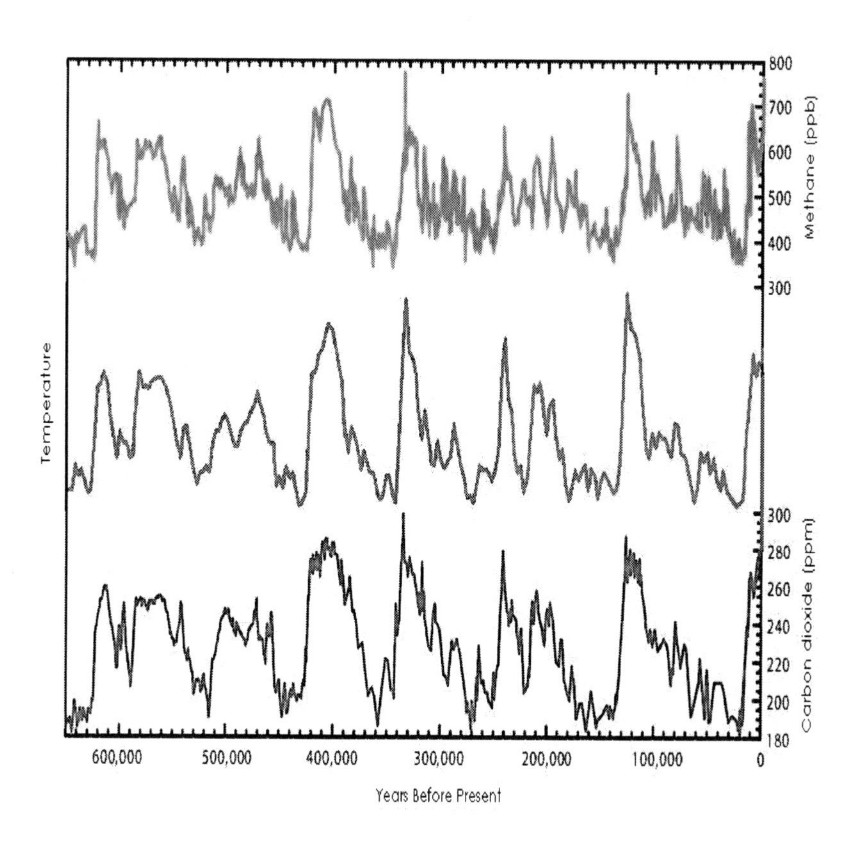

Figure 6.5 – The relationship between atmospheric CO_2, methane and the global temperature is undeniable. If humans are adding vast quantities of CO_2 and methane into the atmosphere what do you think will be the result for the temperature?

---

[i] Ruddiman, W. F. (2001). *Earth's Climate: Past and Future*. W. H. Freeman & Co., New York. Pg 396.
[ii] Cox, P. M. et al. (2000). 'Acceleration of Global Warming Due to Carbon-cycle Feedbacks in a Coupled Climate Model,' *Nature* 408, Pg 184-187.
[iii] Godrej, D. (2001). *The No-Nonsense Guide to Climate Change*. New Internationalist Publications Ltd. Oxford. Pg 21.
[iv] Rosenzweig, C. & Hillel, D. (1995). 'Potential Impacts of Climate Change on Agriculture and Food Supply,' www.gcrio.org/consequences/summer95/agriculture.htmlx, Accessed 17th November 2006.
[v] Kennedy, G. (1999). 'Ozone protection: introduction', *The Planetary Interest*, UCL Press.

Chapter 7 - What Are We Doing Wrong?

An ominous question: what are we doing wrong? There are three steps to answering this question, and the first is to recognise that we are doing wrong in the first place. Atmospheric carbon dioxide is soaring to levels not seen for millions of years and other greenhouse gas emissions are following close behind. Though our sulphate emissions blanket the sky and keep out some solar radiation (creating a small, localised cooling effect) they also trap outgoing radiation (creating a greenhouse effect) as well as causing illness in children and adults, acid rain and so on.

Climatologists have proven that when greenhouse gases get too abundant (or not abundant enough) in the atmosphere, then there are local and global changes to the climate that follow, and sometimes the entire makeup of the natural world is affected. If we then look at our own rising greenhouse gas emissions, and put two and two together, it is more than obvious that we are currently in the fast lane heading towards a big problem.

We are Swarming Consumers

The second step to answering the question is to recognise that the circumstances are changing all the time. Perhaps the biggest issue we can look at is population, because it would be an issue whether we were bad with our environment or good with it, as well as the fact that it is fairly easy to keep a general track on our numbers.

From the dawn of our ancestors, around two million years ago, all the way up to the 1800s the number of humans on the planet grew at a very steady rate. By 1804, therefore, though humans had spread into almost every pocket of habitable land in every continent, the total worldwide population was only just creeping past the 1,000 million mark (less than the present-day population of India). In this initial stage of population growth, the actual rate was incredibly slow. It was only about five thousand years ago that the British Isles were populated for the first time, and it took roughly a thousand years for the worldwide population to double. However, the nineteenth century saw a drastic change in this trend, and it was about 1927 when the population figure was breaking through the 2,000 million people boundary; what had taken around two million years was now repeated in just over a century.

And the rate kept on increasing, passing the 3,000 million mark in 1960, and the 4,000 million mark in 1975. And *still* it keeps booming: with a world population today of just under 6,800 million (though it is hard to be accurate) and increasing by three people every single second.[i] At this rate, with all data taken into account, the population could reach 10,000 million by 2034. If this was any other species on Earth scientists would call it a 'swarming stage' and would be seriously concerned for the well-being of other species and other ecosystems, perhaps even calling for intervention to cull numbers so that population is kept low. For humans that isn't an option, but we are left to sit and ignore it all, as if it isn't a bad thing!

On a planet of finite land and finite resources, a population of 6,000 million is hard enough, but one of 10,000 million is inconceivable. We would have an issue all of its own - in terms of food, water and land shortages – even if there were no issue with greenhouse warming of any sort. But *with* the issue of greenhouse warming, we have a huge problem on our hands. How do we feed, clothe, and house the millions of people of tomorrow (let alone the millions alive today) and provide them with clean water and medicines, without fossil fuels, or deforestation?

Technology is not a bad thing, in fact when it comes to medicine and certain sciences it can only prove good. For example, we would know little of what is in Part 1 of this book were it not for certain advances in technology. We would perhaps still be using leeches and prayer to cure our ailments were it not for medical advances, of which technology has been a critical part. But with technological advances must come *philosophical* advances about how we live in our world.

There are many examples of technology advancing to a point well beyond what our current world-view can deal with. A good example often used is that of nuclear weapons, but to keep within an environmental mindset we can instead look at agriculture. For many millennia people have been using irrigation as a method of boosting their crops and feeding their family but since the early twentieth century, (and the invention of the tractor, the combine harvester and other such mechanical devices), irrigation has grown in scale and magnitude. Farmers can irrigate bigger fields, meaning more land has to be bought or cleared of vegetation, and vast and complex systems of water pipes, pumps and tanks have to be installed.

For an individual farmer this usually turns out to be a benefit, because it means that they can produce more and therefore make more money, but for the environment this turns out to be a problem. We are now diverting more river water than ever before on our planet, leaving rivers drained, often to the point where they no longer contain enough water to reach the sea. When the Soviet Union diverted water away from two major rivers feeding the Aral Sea, little did they suppose that they were taking so much water away that one day the Aral Sea would be drained completely, so that it is today more like the Aral Desert.

The same thing is happening with mining and deforestation. When technology begins to shoot forward at a faster rate than our ethical approach to nature, what we get is strip mines, mountain-topping mining, large-scale deforestation and industrial-scale logging. The leftovers are habitat loss, extinction, pollution, visible landscape scarring and global climate change. And it is the most technologically advanced nations on earth that are causing greenhouse warming. Take a look at how moderately poorer countries contribute to climate change:

Country	**Population in 2004 (millions)**	**CO_2 released (combustion only) in 2004 (million tonnes)**	**CO_2 released per person in 2004 (tonnes)**
Bangladesh	139.22	33.55	0.24
Bolivia	9.10	10.45	1.16
Democratic Rep. of Congo	55.85	2.24	0.04
Ecuador	13.04	21.92	1.68
Guatemala	12.30	10.33	0.84
Morocco	29.82	35.54	1.19
Romania	21.69	91.49	4.22
Vietnam	82.16	78.80	0.96

Figure 7.1 (above) – Carbon dioxide output of nations either too rich to be classified as 'Poor South' or with significantly large populations. *(All data from International Energy Agency. 2006)*

Then compare this with the richest countries in the world:

Country	Population in 2004 (millions)	CO_2 released (combustion only) in 2004 (million tonnes)	CO_2 released per person in 2004 (tonnes)
Peoples Rep.China	1296.16	4732.26	3.65
France	62.18	386.92	6.22
Germany	82.50	848.60	10.29
Israel	6.80	62.21	9.15
Japan	127.69	1214.99	9.52
Russia	143.85	1528.78	10.63
UK	59.84	537.05	8.98
USA	293.95	5799.97	19.73

Figure 7.2 – Carbon dioxide output of some of the richest nations on the planet. *(All data from International Energy Agency. 2006)*

In the first table we either see countries where both population and economic power cannot be called insignificant, and whose status in the world is higher than that of Poor South, *or* we have Poor South countries that have significantly large populations. In the second table we see arguably the eight wealthiest and most consumer driven countries on the planet – regardless of population. The comparison is startling.

First, we can see that a large population doesn't necessarily produce a lot of CO_2 emissions; this can be seen when we compare the population of Bangladesh and its total annual CO_2 emission figure (34 million tonnes) to that of Russia, which has virtually the same amount of people but emits an enormously high amount of CO_2 (1529 million tonnes). Compare also Vietnam and Germany – both with similar populations but with very dissimilar total CO_2 emissions.

Population alone cannot account for climate change while countries are still lacking the technology. But *with* technology climate change can transform even a tiny population into a big player at the CO_2 emission game. Take a look at the figures in the final column of the tables, showing how many tonnes of CO_2 each person in that particular country emits every year. For the nine million Bolivians, the amount each person emits (called 'per capita') is around one tonne per year. For the 30 million Moroccans the per capita emission figure is also one tonne. Even the 'bad-boy' of the table, Romania, has a per capita figure of just 4.2 tonnes.

Now compare this to the nations who have more technology. The 62 million or so French citizens emit over six tonnes each; a similar population in the UK emits just under nine tonnes each. In Russia and Germany this figure goes higher than ten tonnes of CO_2 emitted per person per year, and even a small population of 6.8 million Israelis emit 9.8 tonnes each. This leads us nicely on to the United States.

The United States is different from all others, even in the second table of the 'super-rich-super-polluters.' This is the biggest and most powerful economy on the planet, where the very theology behind consumer-culture starts and ends. As a result, with a fairly large population to give it the extra push, the United States emits 5800 million tonnes of CO_2 every year – 23 per cent more than China but with one fifth of the population. In other words, the average American emits 20 tonnes of CO_2 every year by themselves – if everyone on the planet emitted this same amount, our existence would be supremely compromised.

But why? What makes the US so different from other industrialised nations? Does this chasm between the US and the rest of the world exist because they have more technology than everyone else? No; Japan is one of the world's most technologically advanced nations and its per capita emissions still only come to 9.5 tonnes per year. Is it because the United States has a very different society than the rest of the world? Not really. We see very close mimicking of American society all across the Rich North, especially in places like the UK, Germany and Israel.

The answer lies in a mixture of the above factors. They have a fatal combination of factors working against them: lots of people, knowing little about climate change, with old fashioned carbon-producing technologies, buying lots of energy-consuming items, and virtually no governmental acknowledgement of climate change as a human-created problem.

Technology isn't a bad thing, but when combined with other factors, or when not implemented in areas that really matter, technology can work drastically against the planet.

We Work Against Nature

Agriculture contributes to climate change in various ways. When forests are cleared and farm waste is burnt, carbon dioxide is released into the atmosphere. Methane is released in huge amounts from rice paddies, as well as the rear ends of an increasing cattle population. All nitrogen-based fertilisers create nitrogen oxide. According to the Intergovernmental Panel on Climate Change (IPCC) basic agricultural practices – such as cattle rearing, rice farming and fertilisation – world-wide account for around 20 per cent of all greenhouse gas emissions. The additional destruction of forests to make new cultivated ground for farmers, as is happening in the Amazon Basin and Central Africa, adds a further 14 per cent of greenhouse gas emissions into the atmosphere, since dead trees cannot absorb CO_2.

The effects of mining on nature is never more obvious than a process that is common in many parts of the world and which involves slicing the tops off mountains, cutting out huge holes, filling in river valleys, deleting vast areas of trees, and moving earth from one spot to another – strip mining. Strip mining is surprisingly common in the United States, despite Americans being fully aware of the negative effects it can have on local wildlife, local habitats and local human populations.

In the Appalachian Mountains of West Virginia, mountain-top removals have caused over 400,000 acres to be destroyed, including 1200 miles of streambeds, producing a large scar across the face of the entire United States. And though many people argue that mining, in whatever country, is something that cannot be discarded because it employs many thousands of people and brings in lots of income, take these facts into consideration: in 1948 some 125,000 people worked the mines of West Virginia, today there are around 18,000. 'Mountain topping' demands less labour and employs more machinery, cutting down on costs to the company; mining in America is no longer having having a positive economic and social effect. Mining companies are usually heavily subsidised by the government, meaning that the American public are paying to destroy their own land. The same story can be told across many countries.

In 3000 BCE, around the time humans were only just beginning to arrive, around 85 per cent of land surface in the UK was covered by ancient, semi-natural woodland of such trees as oak, beech, ash and rowan. Today, recent estimates show that only around 1.5 per cent of land is now ancient forest.[ii]

Worldwide it is estimated that around 72 per cent of forests are threatened with logging, 38 per cent by mining and road building, and 20 per cent through agricultural expansion.[iii] Now, if you are sharp you will notice that these figures add up to more than 100 per cent, but that is simply because lots of forests face multiple threats. Figures from 1997 show dramatic loss of forests in nations in every corner of the planet's surface: more than 50 per cent in Bolivia and Brazil, more than 60 per cent in Belize, Bhutan, Columbia, the Congo, Ecuador, Gabon, Indonesia, Nicaragua, Panama, Papua New Guinea and Russia, and more than 80 per cent in Australia, Brunei, Honduras, Malaysia, Sri Lanka and Zaire. In Argentina, Bangladesh, Myanmar, Cameroon, Cambodia, Central African Republic, China, Costa Rica, Guatemala, Ivory Coast, Laos, Mexico, New Zealand, Sweden, Taiwan, the United Kingdom, the United States and Vietnam the amount of natural forest left standing is less than one tenth, and Nigeria, Finland and India have the dishonour of having cut down more than 99 per cent of all their native forests.

Around 2.5 acres of forest are cut every second around the world[iv] – equating to 214,000 acres per day (an area larger than New York City), and 78 million acres per year. This is 121,875 square miles – an area bigger than Poland! Every single year!

When we look at it from a global point of view we see what impact the advances in technology combined with a booming population have had. Humans have always clashed against nature, with the early Mesopotamians, Romans, Greeks and Egyptians deforesting whole regions to build their empires. What happened to the land of milk and honey in Palestine, or the cedar trees in Lebanon? When technology races ahead and people multiply in swarms, deforestation is what you get.

Perhaps the most symbolic place on earth for relentless deforestation is the rainforest of Amazonia. It is the most famous example of deforestation because: (a) it is the biggest tropical forest on the planet; (b) the *rate* of destruction has always been a vocal point in the Rich North conservation movement; (c) it is the home to the richest range of forest life on Earth, with as many as 300 different tree species per hectare. Despite its great biodiversity, and infinite value to us on a worldwide scale, it is still shocking to see that over 75 per cent of Brazil's climate-changing pollution originates from the clearing and burning of its forests, and ultimately ranking Brazil 4th

in the world for greenhouse emissions. Brazil isn't particularly exceptional when it comes to emitting greenhouse gases but, by getting rid of its vast natural wealth, it metaphorically shoves itself into the world's worst climate change offenders.

The small-scale agricultural endeavours of the masses actually have only a moderate input into the downfall of the Amazon. The real culprit lies with large-scale destruction, caused by monocultures[1], plantations, cattle ranches and just good old-fashioned logging - and it is corporations (most Rich Northern-owned) that are behind it all.

Take the example of soya beans. Large parts of lush rainforest are cleared to make way for soya bean plantations, and the trade is often linked to slavery and land grabbing. Because the soya industry is so intense in Brazil, the country is now one of the world's biggest consumers of pesticide, which has terrible consequences for surrounding species and ecosystems. So who is consuming the soya? Brazilians? Vegetarians? No; it is the cattle, chicken and pigs that end up on our supermarket shelves in the Rich North. Deforestation for no good reason.

Loss of the forests accounts for three quarters of all mammals endangered by human civilisation; that 65 per cent of endangered reptiles, 55 per cent of endangered amphibians and 45 per cent of endangered birds are accounted for by deforestation is no consolation. Though rainforests cover only 3.5 per cent of the planet's land surface, they support more than half of all its known organisms. The IUCN assesses that about half of plant species are threatened thanks to human activity, mostly deforestation. Deforestation of the tropical forests forces species such as gorilla and orang-utan towards extinction.

But one victim we always seem to forget about is the human. Deforestation causes endless and life-threatening problems for many different human communities around the world. Before the Holocene many human communities existed within forests, as a part of the forest ecosystem, and they understood the forest, knew the land and found the key to natural survival that most people on the planet today simply don't have. And while other civilisations grew, took up their consumer possessions and chased private wealth, these people remained part of the natural system, rejecting the consumer culture – and many still do.

---

[1] The act of planting a single crop or tree over a wide area which can impact negatively on ecosystems and biodiversity.

Forests remain homes for people all across the planet, living in partnership with nature. As the industrial deforesters encroach on their homes, these communities are forced to move, paid off with a pittance of money to go and join the outside world without a choice. If they reject they become victims of theft (land is taken regardless), their homes are burnt and their families terrorised, and some even fall victim to murder.

For example, a certain wood company presently deforests part of the Central African Republic and the home of the Ba'Aka pygmies. The Ba'Aka have been forced to live in camps at the very edges of the forests like refugees on the edge of a war zone. Their lives are in tatters, their futures uncertain, the injustice is obvious to anyone, yet the international community doesn't even know of their plight. It is easy for people in the Rich North to think that our own governments wouldn't have anything to do with such terrible behaviour but it turns out they are the biggest supporters of logging companies. The World Bank actively encouraged the deforestation of the Gishwati forest in 1967 so that potatoes and cattle farming could take place - consequently costing the Batwa pigmies their home and livelihoods, forcing them to live beyond the frontier of the forest because they cannot go about their traditional hunting behaviour and retrieve food from within.

The whole issue is a deeply moral one: the forcing of human communities to join the outside world and shift off their ancestral land; the purchase and sale of land behind their backs; the fact that they have no choice when loggers move in with no intention of moving out. It's a simple case of the corporation fighting against people, and every time the corporation will win.

So who are we to blame for this whole tragedy? Well, take a look at it from a business point of view and you'll soon realise the answer. For example, up to a third of World Bank funding in Indonesia goes towards deforestation projects but the organisation doesn't care – it actively encourages the Indonesian government to ignore the fact that over 60 per cent of all logging in the country is illegal, and concentrate on its logging exports, its palm oil exports, and its paper pulp production. The World Bank Reforestation Fund is spent on clearing forests, turning wetlands into rice farming, and building plants for paper pulp production – as well as being spent on its state-run aircraft company, national car project, and to prop up its currency.

Environmentalists blame the crisis in the global forests on the huge corporations. Though governments often come up with inadequate laws to try to satisfy our demand for forest protection, it takes more than just a written law to stop the problem – it takes enforcement too. By backing logging corporations, monoculture plantations and the paper industry, governments are actually doing more to *promote* environmental destruction than prevent it. They can even wield their own strength to keep the forests falling – supporting the arrest and harassment of protesters, indigenous communities and anyone else who stands up against deforestation. And I'm not just talking about the governments of Poor Southern countries, who we almost expect to be corrupt and myopic; governments such as the UK, France and US, actively seek to suppress protesters.

This lack of law enforcement ultimately leads to the fact that over half of forests that are cut down are done so illegally – either because the land is not owned by the logging company or because the land is protected by law, but nobody stops loggers going in anyway. In many cases environmentalists have successfully sued logging companies for such acts and judges have granted stays on all cutting in that area. It doesn't matter - loggers just carry on cutting anyway, sometimes even faster, because they know nobody is going to really stop them. The various justice systems on this planet really do nothing to stem illegal deforestation. The systems are often so slow that loggers are in and out of a forest before they have to appear in court, and corruption is so rife in some countries (especially the US) that the penalties dished out in guilty verdicts are but a scratch in the vast profits the companies can make.

In the US, the system is almost completely invulnerable; as soon as the public scores even a small victory against a logging company the government just changes the rules. For example, in 1995, the US Congress passed something known as the Salvage Rider, a bonus piece of legislation stating that if any trees are cut for the sake of "forest health" then they are exempt from other environmental laws. In other words, if a company wanted to wipe out a forest to sell as timber it could do so under the guise of 'forest health' and nobody could do a thing about it! The public had no voice; legal challenges and appeals were prohibited. The government had basically opened up a free-for-all on American forests and extinguished any chance the public had of protesting or changing the laws.

And now we go back to Brazil for one of the most terrible examples for exploitation this planet has ever known. Since its invasion by European settlers hundreds of years ago, Brazil has become the epitome of capitalism. Today, less than one per cent of landowners control 43 per cent of the land, and just six per cent hold 80 per cent of farmland. Transnational corporations alone hold 14.5 million acres of Brazil.[v] Though Brazil is actually the world's second largest exporter of agricultural commodities around 85 million people are clinically undernourished. Eight million acres of Brazil are owned by the largest twenty landowners, made up of senators, ministers and army chiefs, who have whole regions of the forest frontier cleared with logging, and then 'encourage' restless urbanites to go into this newly cleared land and set up agriculture - and this despite less than ten per cent of soils in the Amazon being able to sustain annual food crops.

The global economic system is rigged to favour destruction of habitats, particularly forests. The World Bank and the International Monetary Fund are on the front lines of the war against nature. Both organisations work for the good of multinational corporations, and will do virtually anything to get these corporations in countries all around the world. So, a poor nation needing financial aid to get back on their economic feet, or to allow them to rise out of poverty, will acquire loans from the World Bank or IMF and be told that they must seek investment from multinational companies – the only thing that can lift them out of poverty. So governments seek investment from big companies to give them a financial boost and raise their employment levels.

But what really happens is that the World Bank and IMF force governments to remove their "barriers to trade" or no companies will come in. A 'barrier to trade' is any law that might impact heavily on the operation of big business - this includes labour laws, minimum wage laws, trade union laws and environmental laws. The World Bank actually recommends governments allow more foreign ownership of land, and ignore environmental records of potential investors. Hence, the largest landowners in the world are timber companies.[vi]

Trees are, after all, the best carbon store. A single hectare of tropical forest contains on average between 100 and 250 tonnes of carbon in organic matter (plus more in the soil).[vii] Without forests, not only can individual bird, insect or primate species, survive, but the Earth itself cannot function as an oxygen-rich, warm planet… with H_2O in all three chemical forms. We are fighting a war against the natural world, jeopardising the long-term survival of the planet and its life-forms for our own short-term benefit; a war in which the industrial-scale habitat destruction is just one of many fronts; and all the while we are cutting down the one thing we need above all else to help us in our battle against climate change: trees.

We Choose Poorly From What Is At Our Disposal

Our species is addicted to fossil fuels, highlighted by the fact that we are still languishing in the Age of Oil where our entire economies rise and fall in reflection of oil prices. But exactly how addicted are we to fossil fuels? And how well are we doing in the bid to introduce renewable sources of energy and thus tackle CO_2 emissions? To answer, let us look at some significant facts and figures.

Using the International Energy Agency's 'Key World Statistics 2006' we can look at what is known as Total Primary Energy Supply (TPES)[2]. The chart below illustrates how world TPES has evolved since 1971 to 2004.

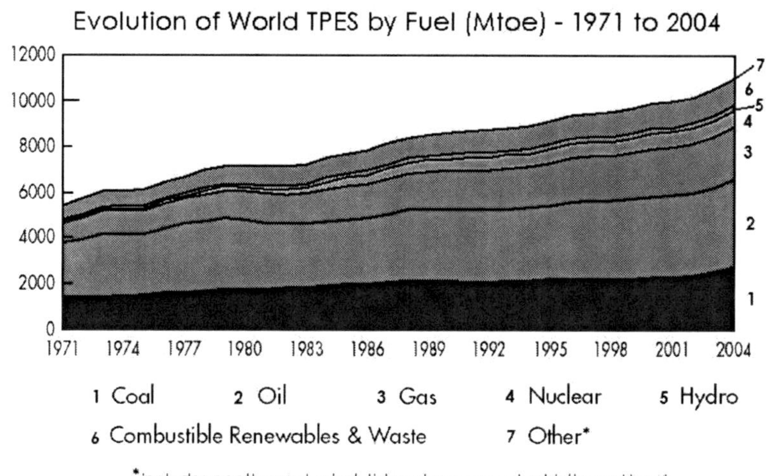

[2] TPES is made up of indigenous production + imports – exports – consumption by ships in international navigation.

Figure 7.3 (above) – The evolution of world Total Primary Energy Supply from 1971 to 2004. Fossil fuels continue to dominate our energy production, regardless of how bad they are for our planet. On this scale, renewable sources of energy (shown as 'other') is invisible. *(Adapted from International Energy Agency. 2006)*

The first thing we should notice is that the two sources showing the biggest growth since 1971, and particularly in recent years, are oil, gas and coal. Furthermore we can see that combustible and non-combustible renewable sources have hardly grown at all (non-combustible renewables include geothermal, wind, solar, tidal etc. and have grown from 0.1 per cent of total energy supply in 1973 to just 0.4 per cent in 2004) and are nothing more than thin lines – even today. It is also worth knowing at this point that, since 1973, the world's total energy demand has grown from 6035 million tonnes oil equivalent (Mtoe) to 11059 Mtoe.

Below are some figures from the OECD countries[3] - the world's most economically developed:

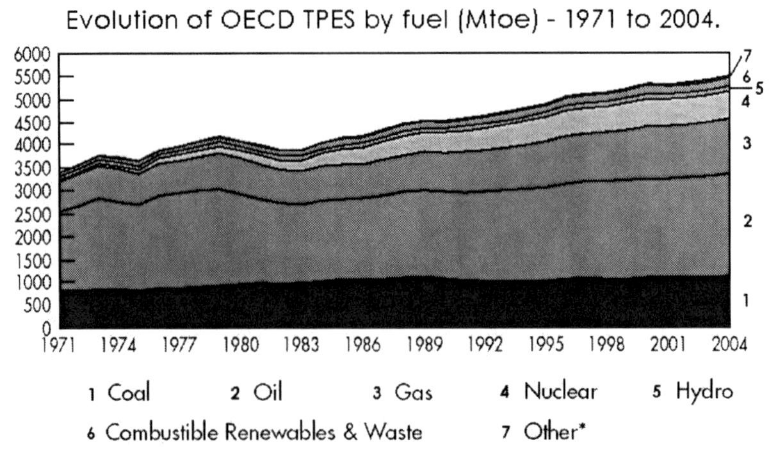

Figure 7.4 – The evolution of TPES of the world's most developed nations exposes the shocking lack of renewable energy sources in use. By far the largest growth has been in fossil fuels. *(Adapted from International Energy Agency. 2006)*

[3] OECD nations: Australia, Austria, Belgium, Canada, the Czech Republic, Denmark, Finland, France, Germany, Greece, Hungary, Iceland, Ireland, Italy, Japan, Korea, Luxembourg, Mexico, the Netherlands, New Zealand, Norway, Poland, Portugal, Slovak Republic, Spain, Sweden, Switzerland, Turkey, the United Kingdom and the United States.

Even in the most economically developed countries, use of fossil fuels is still growing, overshadowing the microscopic growth of renewables; which begs the question, if the richest countries can't bring in renewable energy sources on a scale that makes a difference, then how is the rest of the world to do so? The biggest percentage growth has actually been in nuclear power, but this remains a contestable alternative to fossil fuels.

Since a peak in 1988, coal consumption has been falling gradually, from about 800Mtoe to about 500Mtoe in 2000. Since then there has been resurgence in coal consumption throughout the world, led mostly by the demand of industry, pushing the total towards 650Mtoe in just a few years.

China produces 2226 million tonnes of coal every year, more than double that of the second country down – the US – yet its economic boom has been only fairly recent. There are bigger culprits; not only is the US the third biggest producer of crude oil in the world (7.8 per cent of world total) but it is also the largest importer, bringing in 577 million tonnes of the stuff, which is two and a half times more than the second biggest importer - Japan. The US is also the world's biggest importer of natural gas and the second biggest producer (18 per cent of total).

So what are all these fossil fuels going towards? We can see the direct effect of our growing need for transport fuels when we look at the sources of the oil-refining process (growing demand for plastic containing consumer products also pushes the demand for refined oil). In fact, production of refined oil has grown in every single sector of the market - including aviation fuels and motor gasoline – except heavy fuel oil, which is the only one to experience a decrease in demand. Again the US is exposed as the largest producer of petroleum products, with 22.6 per cent of the world total, and the largest importer, too (97 million tonnes per year).

Looking at world electricity generation we can also see how worse the situation is getting, with the biggest growth in production coming still from fossil fuels.

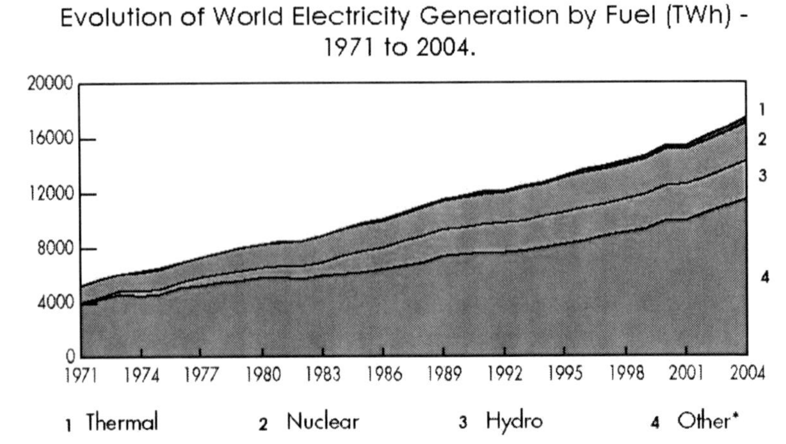

Figure 7.5 – When we look at how electricity has been generated over the last few decades it is clear all of our efforts have been going into fuels that we can burn.

Once again we can see how allergic we have been to renewable sources of electricity production with the 0.7 per cent of 1973 share only increasing to 2.1 per cent by 2004.

COAL	**TWh**[4]	**OIL**	**TWh**	**GAS**	**TWh**
US	*2090*	*US*	*139*	*US*	*732*
China	*1713*	*Japan*	*133*	*Russia*	*421*
India	*461*	*S. Arabia*	*81*	*Japan*	*244*

Figure 7.6 – The top three electricity producers from each fossil fuel. *(All data from International Energy Agency. 2006)*

Fossil fuels, however, may not remain favourable for very much longer. As these finite resources become less abundant, their costs will undoubtedly rise; such rises have already been seen throughout the late twentieth century. The price of crude oil has shot up in the last decade from around $12(US)/barrel in

[4] TWh – TeraWatt hours

1999 to an average of $70/barrel by early 2006. The cost of importing steam coal into the EU and Japan has grown by about $30/tonne over the same period to a high of almost $70/tonne. Natural gas imports via pipeline have also grown since 1999, again to new record highs. With further rises inevitable, the use of fossil fuels will become uneconomical – thus forcing the alternatives forward.

World CO_2 emissions are rising, and one of the biggest reasons for this rise is the growth of fossil fuels in the energy sector. That's right; we are using more and more fossil fuels to provide our energy, instead of less and less like the doctor ordered.

The influence of gas is growing (14.4 per cent of total CO_2 emissions in 1973 to 19.8 per cent in 2004) but it is the increased use of coal that is really pushing emissions higher (40 per cent of CO_2 emissions now come from coal, compared with 34.9 per cent in 1973). This may largely be attributed to the boom in coal production in China, a suggestion enhanced by the country's large growth in CO_2 emissions compared with other parts of the world. This also accounts for China's share of world CO2 emissions increasing from 5.7 percent to 17.9 percent between 1973 and 2004. All other regions of the world have seen their share increase also, except the former USSR, which is now producing less CO2, thanks to the partial collapse of its economic system. The share of emissions from OECD countries may have fallen from 69.5 percent to 48.6 percent, but this does not illustrate a drop in emissions – rather their *increase* in emissions has been slower than other regions.

The Rich North is not turning the tide and becoming sustainable, it is continuing along the path to deeper un-sustainability and the only difference between 1973 and 2004 is that the rest of the world is now on that path too. According to the IEA report, if we carry on down this path the world's TPES could go from 11059 Mtoe today to 16500 Mtoe by 2030. This shows that not only will renewable sources have to be brought in on a huge scale but that they will have to be brought in very soon indeed if they are to curb our greenhouse emissions whilst our energy demand grows like this. As the report predicts, with current policies in place by 2030, renewable sources of energy will still only be meeting 11.8 per cent of our energy demand, whilst 70.4 per cent of our energy will still be met by fossil fuels.

The Nuclear Debate

If our future energy demand grows to the levels experts currently predict, then to wean ourselves off fossil fuels will leave us with a very big energy shortfall. One of the ways to plug this gap would be the use of nuclear power generation. Nuclear power stations have been around for decades and many people think that the only way to give fossil fuels the boot for good would be to switch to a nuclear-powered economy.

The process of generating energy from the nuclear process is quite complex. It involves the fission (pulling apart) of uranium atoms, an act that releases energy. In a fission chain reaction, enough energy is released to create a great deal of heat, which is used to boil water and produce steam, which then drives turbines, rotates generators, and produces an electrical output. The latter half of the process is almost exactly what takes place in a fossil fuel power station (with the production of steam to drive turbines etc.) but without any of the carbon dioxide emissions. The process of nuclear fission itself produces absolutely no greenhouse gas emissions and yet it can generate a lot of energy from a small volume of material. So is this the magical answer we are looking for?

Governments are starting to think so. In places like France, Lithuania and Belgium, over half of energy demand is met with nuclear fission. Governments including the UK, US and Japan have expressed an interest in a new generation of nuclear power to help meet their energy needs or bring down their emissions. In fact, during Tony Blair's term as Prime Minister, he announced several times that nuclear power was going to be key in the UK's energy policy for the next twenty years or so, central to its movement against climate change.

But the *facts* seem to differ from what we commonly hear about nuclear power. First of all, nuclear power generation is not carbon-free. Uranium, the fuel for the reaction process, is like any other fuel in that it lies deep down underground. To mine this fuel and then convert it and enrich it is a complex process, involving input of a lot of external energy. If the uranium is high-grade then the proportion of energy used to produce it is relatively small to the energy that can be produced *by* it – but sadly the vast majority of the world's uranium ore is low-grade. At a grade of 0.1 per cent, 60,000 tonnes of mined rock produces 10,000 tonnes of ore, which produces 10 tonnes of uranium oxide, and only 1 tonne of uranium. At such poor scales, the carbon emissions from nuclear power begin to compare to that of gas production. Even a doubling of nuclear power in the UK would only cut emissions by eight per cent, according to the Campaign for Nuclear Disarmament (CND), because it only tackles electricity production (which accounts for only a third of all emissions) – transport and industry would still be causing problems. It also takes a lot of energy to build and decommission a power station once its lifetime is up.

Nuclear power is also very expensive. The industry is massively subsidised by the public in many countries, including the UK, where the most recent power station built (Sizewell B) cost £3.7billion of taxpayers' money just to build. Decommissioning all of the country's aging power stations will cost taxpayers a further £56billion. The 10 new AP1000 reactors Labour wants built in the UK over the next two decades will cost the tax payer an extra £17billion, since private industry refuses to pay this amount itself.

Perhaps most significant to the nuclear question is the problematic issue of waste. When uranium has been drained of every last bit of its useful radioactivity it remains highly radioactive for thousands of years, and cannot simply be dumped in the sea or in landfill like the rest of our waste. The Committee on Radioactive Waste Management (CoRWM), set up by the UK government, predicted that a new generation of ten nuclear plants will increase high level waste from around 8,000 cubic metres to 39,000 cubic metres – a five-fold increase. British Nuclear Fuels provided their own figure – only a ten per cent increase – to the Science & Technology Committee, and subsequently the government decided that the waste issue was not important enough to affect future plans. This all overlooks the fact, however, that nobody really knows what to do with nuclear fuels once they are spent and many countries rely on shipping their waste abroad rather than bothering with the cleaning up. Roughly five per cent of UK Gross Domestic Product (about £56billion) has been set aside for the costs of clearing up current nuclear waste in the country; few people consider how much more it will cost once new power stations are running.

There is also the added risk of accidents. When an accident struck the nuclear reactor at Chernobyl, in the old Soviet Union, it sparked millions of deaths from radiation poisoning, similar to that experienced following the dropping of hydrogen bombs over Japan during WW2. The entire town was affected, and the radiation spread for thousands of miles over Europe and beyond, thanks to the wind. The health issues are well-documented and range from leukaemia to growth defects and mutations. The effects were even felt on the other side of the planet with both Europeans and Americans experiencing a surge in childhood leukaemia. But how much at risk is there that a similar accident could occur?

Today, nuclear technicians are very strict and careful with the technology at their fingertips, and with more modern systems than those at Chernobyl it is unlikely that another accident could take place – though not impossible. Leaks are reported all the time at plants all over the world, always the result of negligent operation. But are the benefits proposed by nuclear power really worth this risk?

There are risks other than just health issues. According to Michael Meacher, MP, a recent US study carried out, in the post-2001 climate of fear, has revealed that an attack on a nuclear power station could result in 44,000 immediate deaths and a further 500,000 afflicted by terminal illnesses over the long-term.[viii] And what message will nuclear-supporting countries like the UK and US be sending out to countries like Iran, who are being denied uranium enrichment know-how?

According to James Lovelock, the climate change situation is so dire that we have no choice but to take these risks. Energy demands have to be met and fossil fuels have to be removed as soon as possible, and only nuclear can provide centralised large-output electricity. Even a risk of another Chernobyl must be taken, because that is how bad we have let the climate change situation get.

This may well be the case, since the energy crisis and the climate crisis are both converging into disaster. However, consider the fact that nuclear fuel will run out in less than a couple of decades. Taking into account all the energy needed to get rid of past and future wastes, the lack of usable uranium ore left, and assuming that nuclear power continues at roughly its present level, then nuclear power production will use more energy than it produces as early as 2022. Additionally, according to one source, "... if all the world's electricity needs come from nuclear, there [are] only *six* years supply left."[ix]

Finally, consider that climate change is happening *now*. Nuclear power stations can take ten years to build, and even longer to generate electricity. The climate time-bomb is ticking away already, and we simply cannot afford to waste years switching to a source that is both energy-consuming and dangerous. Nuclear power has its good points, but clearly it is too little too late.

The Legacy of This

Over the past two and a half centuries, humankind has been demanding more and more energy to fuel its industry and innovation, leading to bigger populations as nations grow more industrialised - which in turn leads to a greater demand for agriculture, more industry and more air pollution. Expansion of urban areas and agriculture means that you have to chop down trees, and in doing so you are releasing carbon dioxide into the atmosphere, just as you are when you burn fossil fuels, which you need to provide all the energy in the first place. Ultimately what we get is a huge human force in the climate system.

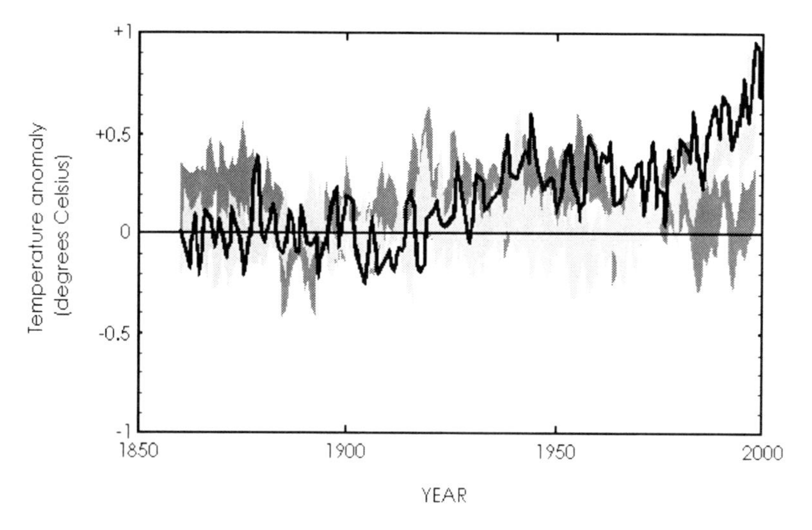

Figure 7.7a – Simulations of global mean surface temperature over the last 150 years can estimate the influence of nature and the influence of humans on the climate. The dark grey area on the graph indicates the state of natural variability from all sources - the light grey area indicates what effect humans should be having on the climate. The black line indicates the actual recorded surface temperature. According to the natural indicators, the global temperature should have been more or less stable - if not dropping – over the last fifty years or so. The black line has actually begun to

diverge from the natural path, following the human influence instead on their climb upwards. In other words, global temperatures rose beyond merely natural influences around the 1970s.

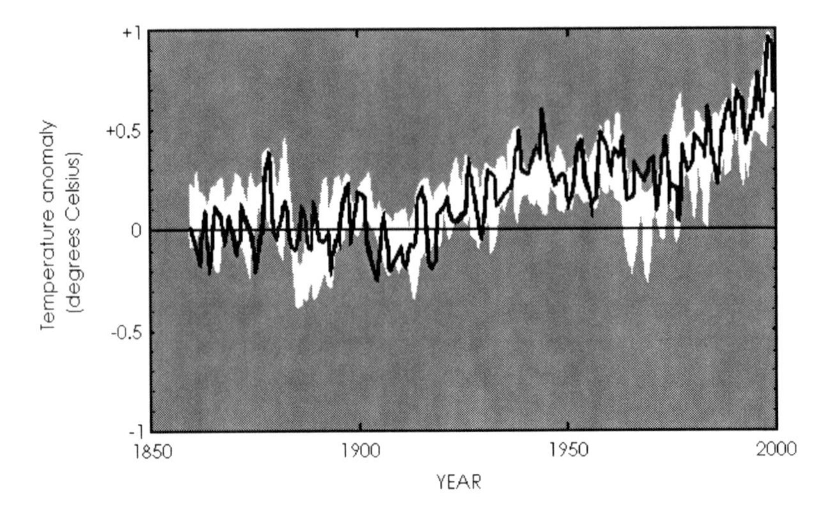

Figure 7.7b – When humans and nature are added together (the two shades of grey in Fig. 7.7a) they produce the above trend (in white). This trend matches the actual recorded temperature change exactly. Together, humans and nature are warming the planet. However, without the human push, the temperatures *would not* be rising.

While the boom in human population seems to be big enough to blame for the increases in CO_2 emissions and other greenhouse gas pollutants, until this point in time population has actually played only a small part in that area. The fact is, populations are booming only in the very poorest of nations, such as the Democratic Republic of Congo, Somalia and Uganda, and remain stable in the richer nations, such as the United Kingdom, France, Australia, and the United States. Poor people cannot afford the luxuries we enjoy here in the Rich North. Therefore, how can a growing population in Burkina Faso be gulping up all the energy and spitting out all the greenhouse gases that make up the boom in these emissions? They cannot: it is the stable populations of places like the United States that are increasing, year after year, their energy consumption and their deforestation, and pushing the world towards climate change.

The future will certainly see a different picture, with booming Poor South populations exacerbating the greenhouse gas problem as they seek the material and social luxuries we have today in the Rich North; this has already begun of course, but *until now* this has not been on any significant scale in comparison to the goings-on of the Rich North. It is easy to point at a booming population and say it is causing climate change, but if that population is incredibly poor, without money enough for medicine and adequate food and water, then the pressure they will exert on climate is only going to be small. It is actually the growing demands of the Rich, and our greater consumption, that is causing the problem.

The lesson we need to learn from the cases of booming economies in highly populated countries such as China and India, but also from the inevitable boom in global population from around 6,500 million in 2005 to 10,000 million by 2035, is that we cannot go ahead with business as usual. There needs to be a huge provision of technology provided to the Poor South to prevent its growing populations exerting the same pressures we placed on the planet when we demanded more energy and more pretty things. This is what we are doing wrong when it comes to the booming population: we are encouraging the poor to find good health and satisfy their basic needs *through consumption* - through the accumulation of products and more resources. For the most part, this encouragement comes from our own consumption. People in poor nations see the material luxuries of the Rich North and want to share in the wealth. They seek to reach the same point we have by following the same path. Even if they reach the same point by way of a different path altogether it is still the same point - the point of over-consumption. As said before, if everybody in the world lived like the average Rich Northerner, we would need several planets' worth of resources to satisfy the demands.

Instead, faced with *more* people on the planet, the obvious thing to do is to have people consume *fewer* resources; to concentrate their wealth on the basic needs and *not* the material luxuries of the consumer culture. And that means that the Rich North has to consume fewer resources, too; maybe, dare I say it, forfeiting the consumer culture altogether.

Since the first human-like creatures to walk on two legs and use stone tools appeared about four million years ago, human history has seen some impressive accomplishments. So good have these accomplishments been that many people alive today are masters of mathematics, physics, astronomy, chemistry and biology, more so than they are of hunting or farming. Never before in the history of the Earth has a life form 'advanced' to the point that the acts of survival – like eating, drinking and finding shelter – are all ready made, packaged or produced on a factory line. Even the relatively 'old' technology of the circuit board is so unbelievably complex that it is a wonder that we have harnessed this technology at all, considering what a combination of elements it is.

But while one part of human history has been accelerating towards greater technology and wealth accumulation, the other half has been decelerating into intense and unbearable poverty. So too has the environment become impoverished: pollution builds year on year, biodiversity takes blow after blow as human society builds, and the most vital key of all – the climate – has become as fragile as a cardboard box in a hurricane. Human 'growth,' both in terms of population and development, has side-effects in itself and the world around it. Our technological advances, combined with our insatiable appetite for consumer items, wealth accumulation and energy, enhanced by our booming population, combines to make the human species a force of nature all of our own.

The modern environmental movement was born during the mid-twentieth century, at a time when changes to the environment were starting to occur on larger and larger scales. Since then, humans have learned more about the environment and how affecting one part of it will have repercussions in the other parts too. I suppose the real answer to the question of 'what are we doing wrong?' lies with the generations of people who have been taught about climate change, and other environmental issues, but did not learn the lesson. We know that deforestation, for example, causes numerous problems, not only with the climate but with ecosystems, with extinction and with ethical concerns – yet we do it today more than ever before. We know that public transport is better for the planet in a many number of ways, but yet there are more cars on the world's roads – more on the roads of the Rich North too – than at any point in history. We know what we are doing wrong and yet we still do it, more today than ever before. What are we doing wrong: everything!

[i] Des Jardins, J. R. (2001). Environmental Ethics: An Introduction to Environmental Philosophy, 3rd Edition. Wadsworth Group, Belmont, CA. Pg 67-68.

[ii] http://www.futureforests.org, Accessed 28th May 2006.

[iii] Jensen, D. & Draffan, G. (2003). Strangely Like War: The Global Assault on Forests. Chelsea Green Publishing Company, Vermont. Pg 111

[iv] Ibid. Pg 16.

[v] Ibid. Pg 91-92.

[vi] Ibid. Pg 128.

[vii] Godrej, D. (2001). The No-Nonsense Guide to Climate Change. New Internationalist Publications Ltd. Oxford. Pg 27.

[viii] Crumpton, N. (2006). 'Nuclear a No-Go Area', the Big Issue in the North, July 10th 2006, Pg. 11.

[ix] Fleming, D. (2006). 'The Lean Guide to Nuclear Energy'. From wwww.theleaneconomyconnection.net/book.htm, Accessed 12th February 2007.

Chapter 8 - The Discovery of Global Warming

By now we are fully aware of the how the climate on Earth is not constant, but a set of actions and reactions – a balance of cause and effect, change and response. We know that climate changes occur naturally over millennial, tectonic and orbital time scales and that these changes vary from slight to extreme. We know that temperature increases on a planetary scale cause higher sea levels, denser forests and larger deserts, whilst temperature decreases generally cause lower sea levels and more polar ice.

We know why this is the way it is, and why it can be no other way. We even know that the most important element of the climate system is the greenhouse effect, caused by atmospheric gases like carbon dioxide, water vapour and methane. Finally we are aware that human beings have begun to add greenhouse gases to those already in the atmosphere. But knowing all this, do we yet fully understand the consequences?

Let me go all the way back to a metaphor I used near the beginning of the book and our image of an egg being balanced on its pointiest end; the last thing needed is for someone to come along and start applying pressure to one side of the egg. This is what human society is doing; it is applying one-way pressure to a system that relies on balance and equilibrium. The result of this can only be unnatural and damaging.

Something Strange

The theory of climate change is not as new as one would think. In fact, it was 1827 when the idea of greenhouse warming came to a French scientist by the name of Joseph Fourier. Fourier was struck by the question of why Earth's atmosphere never seemed to warm continuously, despite being under constant bombardment from the Sun's radiation. He postulated that the Earth's warm surface emitted infrared radiation, invisible to the eye, which carried the heat away into space. But, after calculating this effect, he came out with a temperature well below what the actual Earth is.

Fourier realised that *some* of the heat *must* remain inside the Earth's atmosphere, otherwise it would be far colder than it actually is. The only thing that could prevent some radiation from leaving the Earth, whilst letting through the majority of it, was the atmosphere itself. He came up with the explanation of a box covered by a glass lid: the interior of the box warms when sunlight enters, but the heat cannot escape. His explanation was accepted because it seemed quite plausible, and the idea of a 'greenhouse effect' lasts until this day. Unfortunately, held back by a lack of technology, there were few practical things Fourier could do to take his idea further.

It was British scientist John Tyndall who, in 1859, decided to test the then recognised fact that all gases are transparent to infrared radiation. After long hours in his laboratory, Tyndall concluded that the most abundant gases in the atmosphere (nitrogen and oxygen) were indeed transparent to infrared. He was almost ready to give up when he came up with the idea of trying coal gas – an industrial gas consisting mostly of methane, and used in the common gas lights. He discovered that, in terms of heat, the gas was very opaque indeed, and after going on to try other gases, he also came across CO_2 and marked down its opaqueness; thus was the discovery of 'greenhouse gases'.

Out of all these opaque gases, only CO_2 existed in any significant amount in the atmosphere, and Tyndall saw how it could cause warming.

Tyndall was not really looking for 'opaque' atmospheric gases with the idea of deepening Fourier's theories. Instead he wanted to tackle the great controversy of his time – the prehistoric Ice Age – that was buzzing around the scientific community. Geological evidence from the United States and Northern Europe appeared to suggest that these regions had been covered by ice sheets the size of continents, and more than a mile deep in places. In Tyndall's day, this incredible idea was only just beginning to be accepted.

In the 1870s, another Briton picked up on something fundamental to climate science. The geologist, James Croll, noted that an ice-covered region would reflect most of the sunlight back into space and would remain cool, whilst soils and vegetation would continue to feel the sun's warmth. Once an area has cooled down, Croll argued, the wind patterns would change, and this would in turn alter the currents of oceans; perhaps resulting in greater heat loss across the entire region. So, once something had started an ice age, the cycle would become self-sustaining. The complexity was well beyond technological reach at this time, but the idea of 'feedback' had been well and truly established.

In 1896, a Swedish scientist took on the challenge, and, not wanting to forget the value of water vapour to the climate system, he began to look at carbon dioxide more closely. After months and months of numerical computations, well before the age of calculators and computers, Svante Arrhenius came up with some figures that he published confidently. He had yet to prove that variation in CO_2 would change climate, but had come up with firm, reliable scenarios of the results if CO_2 variation *could* cause climate change. According to his theory, cutting atmospheric CO_2 in half would cool the planet by almost 5°C. When he considered the effect of 'feedback', he concluded that, over time, snow would accumulate and start reflecting sunlight; this may be enough to begin an ice age.

Arrhenius calculated that doubling the atmospheric carbon dioxide could result in a 6°C rise in mean global temperature[i] (a figure almost perfectly in tune with the IPPC's prediction for a doubling of CO_2 – 5.8°). This came after a consultation with friend Arvid Högbom, who had estimates for how CO_2 cycles through natural geochemical processes – oceans, volcanoes, vegetation etc. It occurred to Högbom that he should try to calculate the volume of CO_2 emitted by industrial processes such as factories, which he could clearly see were having some sort of impact on the atmosphere from all the black soot problems that Western Europe was having around that time. To his surprise, he found that industrial emissions were adding CO_2 to the atmosphere at the same rate as nature was adding and removing them. Although in 1896 the amount added by humans was nothing compared with the amount stored in the atmosphere already he realised that over a long enough period of time, the culmination would begin to have an effect.

The Swedes were not as alarmed by this idea as one expects they might have been. First, the idea of warmth for a couple of Scandinavians is not entirely unwelcome, and secondly, Arrhenius had made an important calculation. He worked out that any significant warming of the planet, caused by humans doubling atmospheric CO_2 levels, would take a few thousand years, by which time human technological advances would solve any problems. But this important calculation turned out to be horrendously wrong.

During the late nineteenth century, the world was populated with no more than 1000 million people, most of them peasants without any demands other than what they needed to survive. Arrhenius and other scholars had not accounted for the changing world around them. His calculations were based on consistent industrial pollution and a static world population – two things that were about to experience a mighty booming growth in the following decades. As the population kept growing wildly, and with the consumption of resources and fossil fuels taking off, the estimate of several thousand years was about to come abruptly closer to the modern day than anyone would imagine.

As the twentieth century went from infancy to young-adulthood, the idea of anthropogenic climate change was rebuffed as a series of other theories popped up to snub Arrhenius and others like him. They argued that there was simply too much carbon contained within the earth itself for the tiny atmospheric concentration to hold significance. Furthermore, it was generally upheld that the natural processes that go on in and around Earth were far more complex than anyone, including Arrhenius, could understand. If a change happened in the atmosphere with any significance, some natural reaction would take place to neutralise it and restore order; this was simply the way of things.

These objections gained so much support because they conformed to a view that was almost universally accepted at the time: the Balance of Nature. In this view, human actions were so trivial compared with the sheer scale of the planet that no matter what we did, Nature would maintain her dignity and grace, remaining consistent to the end. This view was rooted deep within most cultures and traditional religious faiths, dictating that God had set the rules of everything in the universe – from the ocean currents to the orbits of the planets. God had created such beauty, majesty and flawlessness, that man would be a fool to presume he could alter it.

The slow advance and retreat of ancient glaciers in the climate record fitted snugly within the 'uniformitarian principle,' – the idea that everything in nature remained uniform and static. How else could something be studied scientifically if it did not always use the same set of rules?

This was a period when science had been hacking away at religious teaching for many decades, and many were trying to establish middle ground that both could settle on. Sanctimonious science and religious scholars often settled the newly forming arguments concerning the natural world with the contention that everything happened by natural process, in a world governed by a steadfast, God-given order.

Science, by definition, cannot let an unanswered question lie, and throughout the early decades of the twentieth century, theories about the cause of ice ages were emerging frantically. As R. Weart points out, "the possibilities spanned half a dozen different sciences." The most acclaimed idea was in the field of geology, and suggested that climate changes were caused by tectonic processes, such as the formation of island chains that might have cut off warm ocean currents, and the uplifting of tall mountain ranges may have blocked prevailing winds. Such explanations didn't seem to fit, as geological movements of this type often take millions of years to occur, and a further theory about volcanic discharge left in the atmosphere became frontline. Perhaps spells of long and large volcanic eruptions had occurred prior to the ice ages, blocking out some of the sunlight and cooling surface temperatures in regions.

Some scientists rejected these theories by supporting the idea of power in the oceans. It was already known that most of the planet's gases were dissolved in the oceans, and that the top few metres of the ocean held more thermal energy than the entire atmosphere. American T. C. Chamberlain found: "the battle between temperature and [ocean] salinity is a close one... no profound change is necessary to turn the balance" By this precedence, the rate and the place that the salty ocean waters sank would determine the climate uniformity. The explanation was subtle, and largely ignored by experts. Still more hypotheses were churned out.

Since the times of the ancient Greek civilisation, ordinary people - as well as scholars – had speculated about the possibility that their cutting down of woodland and clearing of other vegetations might have had some effect on the climate. It was only common sense, especially to rural farming communities, that shifting vegetation cover from forest to field will affect rainfall patterns. But, meteorologists failed to see any historic effect of land clearance on climate from the reams of data they had available and declared that the biosphere had little, if any, impact on climates.

However, the idea stuck around, and it was Russian biochemist Vladimir Vernadsky who gave it depth. Vernadsky worked during the First World War mobilizing the production of industry, and from this viewpoint he was able to see that the volume of materials produced by humans was reaching geological proportions. He began analysing biochemical processes and concluded that the major atmospheric gases of nitrogen, oxygen and CO_2 were there as a result of living organisms. The 1920s saw him publish several works arguing that the entities making up the biosphere constituted a force for reshaping the planet, rivalling those of any physical nature. His visionary idea that human innovation had made our species a geological force of our own was outstanding by far, but it got read by few and was largely unsuccessful.

Following up on another theory by James Croll, Serbian engineer Milutin Milankovitch came up with the idea that the wibble of Earth's axis would result in periods where high-latitudinal zones would receive much less sunlight than usual. He said that a lack of heat in summer would fail to melt accumulated winter ice, and this would build up over the years, slowly leading to an ice age; as the axis gradually tipped the Earth back towards the sun, the ice age would gradually melt away in accordance.

But his calculation failed to match the timings of the four ice ages that humanity had managed to gather data on by this time and, yet again, another theory fell behind in the race to explain ice ages.

Finally, in 1938, a man named Guy Stewart Callendar stood before the Royal Meteorological Society in London and proclaimed that their accepted ideas of a Balance of Nature were false. Callender was no meteorologist, and only managed to study it as a hobby when he wasn't working on steam power, but he was the man who would boldly declare - in front of an audience of bemused onlookers - a theory that would have a resounding affect from that day to this. He had pieced together the theories of Fourier, Tyndall and Arrhenius and, after his own data confirmed his suspicion that global climate was indeed warming, he singled out CO_2 as the cause. He finished by stating his belief that the source of this additional CO_2 was the growing emissions of industry worldwide: humans.

Unfortunately, Callender calculated that the average global temperature would not rise by more than one degree by the 22nd century! He even speculated that global warming would aid crop growth and be a good thing for humanity. Even so, his idea went more or less unheard and, for the next few decades, the idea that human CO_2 emissions were causing global warming was set aside to gather dust.

Climate Change makes a comeback

In 1956, Gilbert Plass reawakened the hibernated topic of CO_2 and climate change. After extensive calculations he was able to show that adding or subtracting carbon from the atmosphere *would* change the amount of solar radiation escaping from the atmosphere. Plass not only pointed out that CO_2 would affect the greenhouse effect, but his calculations also stated that the average global temperature would rise at the rate of 1.1°C per century.

Barely a year later, Roger Revelle and Hans E. Suess completed their research into the levels of carbon absorption in the oceans. Prior debate on the topic had concluded that, even if humans *were* adding high levels of carbon to the atmosphere, it was only logical to believe that the hugely capacitated oceans would soak it all up. Revelle proved otherwise - realising that the mix of chemicals in seawater creates something of a buffering effect, stabilizing the water's acidity, and altering the balance through a chain of chemical reactions. Whilst most of the CO_2 added to the atmosphere *was* being absorbed by the ocean surface, the majority of the molecules were evaporated back into the air. Calculations showed that the oceans were not actually absorbing much of the gas at all in total, and whatever humanity added to the atmosphere would take thousands of years to disappear.

However, this conclusion was a terrible underestimate. Revelle had joined the long list of theorists that had failed to see, and account for, the explosion of population and material consumption that began during the industrial revolution, and continues even until this day. The exponential growth rate of these two factors had seen a quadrupling of both population and energy use per person between 1900 and 2000 – resulting in a 16 fold increase in the rate of world CO_2 emission. Just before sending his paper off for publication, Revelle quickly added a final aside, concluding that "Human beings are now carrying out a large scale geophysical experiment of a kind that could not have happened in the past, nor could be reproduced in the future."

Revelle and Suess' work was (you've guessed it!) largely ignored by most, but finally some were beginning to see the significance of a human-CO_2-

global warming link. The pair was unperturbed with the lack of enthusiasm, and applied for a modest amount of money from the newly formed International Geophysical Year of 1957-58 (IGY) which had been set up by a disgruntled band of scientists, aiming to encourage worldwide co-operation amongst the various geophysics disciplines. The IGY had become highly valued by governments, who found that using this guise they could collect global geophysical data of great military value. Hence, the organisation received massive funding from many nations, particularly the ones climbing down each others throats during the height of the Cold War.

Revelle's request for money was to fund the finding of a "snapshot" of baseline CO_2 values on a global distribution; he proposed that after a decade or two, someone else could come along, take another snapshot, and then decide whether or not the levels of the gas had risen. Because of its simplicity, the funds were granted and a young geochemist was hired to undertake the task: Charles David Keeling.

A young Californian with an eye for adventure, Keeling had other ideas and proposed something more extensively accurate. Somehow, despite the apparent uselessness of his objective, Keeling managed to persuade some key officials to part with their money and pay for his expensively accurate machines for recording carbon dioxide. He set one of the instruments atop the volcanic peak of Mauna Loa in Hawaii, almost isolated by thousands of miles of pristine ocean, and probably one of the best sites in the world to study the untouched atmosphere. Another went to the even more remote and immaculately fresh Antarctic.

Keeling was meticulously precise with his readings, managing to capture a surprisingly accurate baseline number for the level of CO_2 in the atmosphere. Within just a year, the Antarctic data was already showing a rise, much to the astonishment of Keeling and Revelle. Following the second year of study, data was appearing to show that the baseline of atmospheric CO_2 level had risen; the rate of this rise was exactly what would be expected if the oceans were not absorbing the majority of human emissions.

By the 1950s, human beings were dumping 5,000 million tonnes of greenhouse gases into the atmosphere per year. Keeling continued to take climate readings from Mauna Loa in Hawaii as new discoveries relating to

CO_2 and climate were taking place all around him. In 1958, for example, telescopic studies showed a greenhouse effect created temperatures on Venus far above the boiling point of water. Additionally, strong evidence was emerging that past global temperature changes (of the last 600,000 years or so) were linked to the level of CO_2 and Methane.

Keeling's data continued to show annual increases in atmospheric carbon dioxide and his measurements continued well after its intended expiry date; the temporary post Keeling had taken in 1957 turned into a career of a lifetime. It didn't take many years for 'Keeling's Curve' (illustrated in Figure 8.1) to catch on and, from that point, all sorts of scientific journals were citing it as the quintessential icon of a greenhouse effect.

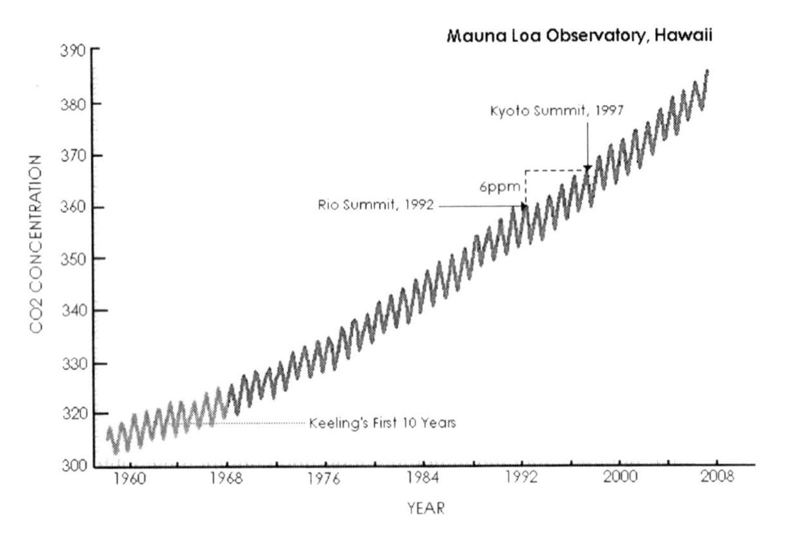

Figure 8.1 – Observations of atmospheric CO2 concentrations far out in the Pacific show increases every year since the study began in 1957. Notice that the 6ppm increase of the first ten years of Keeling's study took just the five year period between the 1992 Rio Earth Summit and the 1997 Kyoto Summit. Not only is the concentration increasing but the *rate* of concentration is also increasing.

Author, R. Weart, describes this point in the story as the "capstone on the structure built by Tyndall, Arrhenius, Callender, Plass, and Revelle and Suess." As far as most were concerned, this wasn't proof of global warming, and technically it wasn't *the* discovery, but this structure built by the gentlemen above certainly stood as a towering monument to the possibility – the theory – of global warming; lacking only the foundation of concrete proof.

Storm Clouds Gather Above

The conference on 'Causes of Climate Change' held in Boulder, Colorado, in August 1965, was headed by Roger Revelle. The summit was barely a blip on the international scientific calendar at the time, but with hindsight it is possible to see that this meeting was nothing less than a turning point. Although intended to look at the many arguments concerning the causes of ice ages, the conference ended up exploding into hot debate surrounding the issue of future climate prediction. One of the most fiercely contested questions was whether human industry really was a force big enough to rival that of orbital and tectonic climate changes.

Revelle took the initiative and went to great lengths to talk to journalists and government officials about global warming. He even picked up on the idea that climate changes might have befallen some civilisations during ancient times, and warned that Texas and Southern California could become deserts within the next hundred years.

Maurice Ewing and William Donn took the idea that climate shifts could happen *rapidly* (in 'science-talk' this means a couple thousand years or so) to realistic heights. Their suggestion that a cyclical feedback system – warm polar sea, leading to more evaporation and more snow, leading to ice sheet growth, leading to sea level fall, cooling of warm currents, less evaporation, melting of continental ice sheets, sea rise, oceanic ice sheets melting… and so on - could happen within a matter of centuries, supported emerging evidence that the last great ice age had not ended quite so smoothly as everyone had once presumed.

The rise of GCM (General Circulation Models) was made possible by the availability of computers - still a long way away from a modern home PC, often taking up several walls of a large room. Scientists now began weather prediction with more precision, and it was this gain in accuracy that suddenly revealed the climate system for what it truly was: a complex system, of immense fragility. If a number was incorrectly inputted into a GCM by the operator, the whole system changed; even if that number was out by a tiny discrepancy, several decimal places down the pecking order, large changes would occur. Whilst many blamed the poor abilities of computers, some questioned whether the jumps revealed something important about the real climate out there; perhaps a minor alteration would trigger a major shift.

This problem was taken up by Edward Lorenz from the Massachusetts Institute of Technology, who had discovered just how fragile the climate system was through a systematic mistake in his inputting of data into his own forecasting model. A tiny alteration to data would mean the difference between a heavy storm and a cloudy day. He brought his findings to the Boulder meeting of 1965. The conclusion of the conference contradicted the accepted, traditional beliefs of meteorological science for the previous two hundred years: the climate was not a single entity, changed by a single force – it now appeared to be influenced by all sorts of external and internal factors. Changes were, in fact, the result of numerous acting forces, all complexly interactive. It became apparent to those present that studies of global climate didn't just lie in the domain of meteorology, but instead it spanned several sciences - all needing to work together to produce results worthy of its grandeur.

Convincing the World

The idea that mankind could be affecting the climate through industrial (and household) emissions was not beyond credibility. In fact, the rumour about global warming fits in quite well with the growing environmental movement, flowering alongside the anti-Vietnam war sentiment that was spreading around the world. By the early 1970s, the theories would not go away, and a conference was set up completely dedicated to the "Study of Man's Impact on Climate." The Stockholm conference of 1971 concluded that the global climate *could* shift dangerously within the next hundred years, as a result of "man's activities."

Savage droughts rocked agriculture in the Soviet Union less than a year later, causing widespread disruption to world food markets, and, in the US Midwest, precipitation fell to a noticeable low. Around the same time, the annual monsoon failed to arrive in India, with dire consequences for the population dependent on it. At the pinnacle of this dry season, the consistent drought that had ravaged the African Sahel for years reached a dramatic peak in 1972, killing hundreds of thousands of people in the region, starving millions more, and provoking mass migrations across the continent. Suddenly, there were real and dangerous consequences to a changing climate, and the television images seen throughout the Rich North brought the reality of death right into people's living rooms.

One study published in 1969 pointed to a similarity between the temperature patterns in Greenland and Antarctica - where one went up so did the other – suggesting that both hemispheres experience the same climate changes. The laborious yet fragile study of the tiny bubbles of trapped atmosphere lying in the ice paid off by providing accurate data of frequent intervals throughout geologic time, more than 100,000 years old.

Funnily enough, it was a 100,000-year period that stood out in both deep-sea core samples and a simple layer pattern lying in some Czechoslovakian soil, spotted by a man named George Kukla. Since the theory couldn't be tested by traditional radiocarbon dating, because its rate of decay prevents any from existing more than a few tens of thousands of years, it was left unto Nicholas Shackleton to devise a system using the new technique of radio-active potassium, and thus pin down some dates for the sediment samples found on the sea-floor. Shackleton's analysis confirmed that there had not been just four major ice ages, but dozens of them, each ice age waxing and waning over cycles of 20,000 and 40,000 years, as well as the dominant 100,000 year cycle. Milankovitch's orbital theory had been verified by clumps of dirt from the sea floor, but the magnitude of the situation was not lost on climate scientists - the vast majority of whom were certain that the Earth's orbit drove the climate over the very long term. Extrapolation of the curves (predicting the near future by carrying on the apparent trend: up or down) appeared to show that the world was actually due for another ice age, in contrast to the previous theories surrounding 'global warming'.

The analyses of the Earth's orbital cycles were pointing to an apparent approach to the end of the planet's current warm epoch. The focus then shifted from a fear of global warming to a fear of a new ice age, turning the study of future climate change on its head.

And so, as the 1970s went into full swing, more questions than answers surrounded the field of climate science, but scientists were beginning to assemble a coherent picture of how and why the world worked the way that it appeared to. There was a sudden realisation that the great unknown future may be dictated by a harsh and unprecedented climate shift, and that human 'progress' was probably the cause of it all. Should science call for immediate action to counter this potential threat? The consensus became polarised: some built up the threat, others played it down. As one expert remarked: "I guess I'm rather conservative - I really would like to see a better integration of knowledge and better data before I would be willing to play a role in saying something political about this." Another simply replied: "To do nothing when the situation is changing very rapidly is not a conservative thing to do!"

Throughout the 1970s, fear about global warming hitting humankind within the next century accelerated, and a few scientists at the 'concerned' end of the spectrum no longer waited for science journalists to pick out the stories from the research papers – instead they addressed the public directly, publishing books about the matter. By the end of the decade about a third of Americans had at least heard about the 'greenhouse effect.' For most who *had* heard of it, global warming was no longer the consequence of nuclear testing or smog – it was firmly understood that the source was increasing CO_2 emissions, attributed to human industrial activities. Even so, it should be pointed out that by the start of the 1980s, resolving global warming was far down the list of anyone's priorities.

The sceptical backlash was strong, and the thirty-year long cooling trend between the 1940s and 1970s fuelled their fires. How could humans be warming the planet if the average temperature had been decreasing for so long? The answer came not far into the 1980s, when two separate research teams in New York and East Anglia discovered that the three warmest years in the last 134-years (that was the length of their records) occurred during the same decade: the eighties. Predictions materialised stating that with the present rate of CO_2 accumulation, carbon dioxide warming should "emerge from the noise level of natural climatic variability" by the year 2000. Furthermore, scientists in New Zealand had pointed out that the cooling trend over the preceding thirty years or so had not been global – the temperatures in the Southern Hemisphere had been rising all the time. The Northern Hemisphere had cooled because of industrial haze, and since the majority of global industry and population live in the North, the Southern Hemisphere had not been affected by the smog's cooling effect.

In 1985 it was time for Wallace Broecker to discover, or expand on, the idea of a global ocean circulation, controlled and driven by the great conveyor belt. The discovery of the rapid climate cooling that had occurred at the end of the last ice age (as the world warmed), and caused by the melting of Lake Agassiz, confirmed the sensitivity of the ocean conveyor to salinity, and the massive impact an altered conveyor speed could have globally. Broecker saw this potential for a swift and spectacular climate shift, and was on the frontlines when it came to taking this news to the public. The climate, Broecker said, was like an erratic beast, and we were poking it with a sharp stick.

The conservative backlash to climate change theories gained substantial ground in the early 1980s when the Republican candidate, and ex-actor, Ronald Reagan was elected President of the United States – bringing with him an administration of sceptical right-wingers, contemptuous of global warming. In fact, traditional conservatism called for a 'business-as-usual' attitude towards the economy, and environmental issues were often clumped together and discarded as liberal nonsense. The new National Climate Program Office suddenly found itself serving as "an outpost in enemy territory," as Robert Fleagle puts it. The new administration set about hacking away at CO_2 research program funding, and discarded an Environmental Protection Agency report on the issue.

However, despite fending off environmentalists for years, the Reagan White House suffered a defeat when, in 1985, a British team of scientists discovered the 'hole' in the ozone layer over Antarctica. Reagan was forced, by public outcry, to sign up to the 1987 Montreal Agreement, formally pledging to ban CFCs from all products, not just air canisters.

The environmentalist movement took climate change as its main fighting point after the success of the 'no CFC' campaign, and, in response, the United Nations and the World Meteorological Organisation decided to create the 'Intergovernmental Panel on Climate Change' (IPCC), a hybrid organisation of science and politics, in 1988.

It became noticeable by the end of the decade that those scientists who rejected global warming scenarios were not talking to sources such as scientific publications (where statements are reviewed by other experts prior to print) but only at events funded by corporations, industry or conservatives - or in business-orientated media like the Wall Street Journal. As a result, criticism mounted against climate sceptics and the basis of their arguments was lost in the squabbling. The media exacerbated the two opposing camps and a brawl broke out encompassing both science and politics. At the centre of the conflict, the general public were simply confused: having been warned of impending doom by one set of people, and then being told to ignore it by another, the feeling of 'to-hell-with-you-all' generally numbed all care for the issue, especially at a time when the West was celebrating the dismantling of the Berlin Wall and talking about the Gulf War at the dinner table.

The Slow Boat to Redemption

In 1966, Albert Gore Jr. was the son of a senator and a Harvard undergraduate. Whilst at university, Gore attended a lecture given by Roger Revelle discussing the future of the planet; Gore was amazed by the sight of Keeling's curve of atmospheric CO_2, at this point eight years old. He was especially impressed by the idea that human beings could directly orchestrate their own downfall, as well as that of the Earth as a whole.

As a Representative in Congress in 1981, Gore was in a better position than any other politician to give greenhouse warming a political foundation. He tried to embarrass the Reagan administration with a Congressional hearing on their proposed cuts to CO_2 funding, but only succeeded in bringing the topic to the alert of the mainstream media when testimonies by Revelle and Schneider were given. However, by 1994 a new democratic President of the United States had been elected, and Al Gore was sitting in the seat marked 'Vice-President'. One of his first actions was to urge the new President, Bill Clinton, to officially commit the nation to the targets set by the 1992 United Nations Conference on Environment and Development in Brazil – commonly known as the Rio 'Earth Summit'. The Agreement established in Rio was signed by more than 150 nations, and obligated each of them to commit their economies to targets aiming to cut carbon dioxide emissions over the next couple of decades.

The 1997 Kyoto conference in Japan was a festival of 6000 official delegates, thousands of other stakeholders, such as representatives of environmental organisations and industry, and a horde of journalists. From the beginning it was clear that most parties had already drawn their line in the sand and were ready to defend it at all costs. For the nations of OPEC (Organisation of the Petroleum Exporting Countries) any tough emissions targets would hit them hardest so they refused to accept targets that were more than just symbolic. European opinion favoured stronger emission cuts but some members, such as Norway and Iceland, argued against cuts because of their own large reserves of fossil fuels still untapped.

The US, however, refused to be penalised unless everyone else was too, arguing that Poor South countries would be given an unfair economic advantage unless they faced equal carbon cuts as the Rich Northern states. This line had been created back in the US Congress, with a resolution co-sponsored by Democratic Senator Robert Byrd and Republican Senator Chuck Hagel. They argued that, unless developing nations had tough targets set on greenhouse gas emissions the same as those set on the US, as soon as the Kyoto agreement was ratified, jobs and money would flow out of the country and to the poor like water down a plughole. They created an image of an American economy starved of employment and industrial competitiveness and the story worked its magic in Congress and amongst the public. The pair went round making lots of noise just as Kyoto was coming around, and played a big part in forcing the agreed targets to fall well below what was required to stabilize the rapidly changing climate. However, maybe it wasn't purely a patriotic concern for the economy that spurred them on in their cause. Maybe it was the fact that the pair had made almost $300,000 collectively in political contributions from fossil-fuel related industries in 1996 alone.[ii]

The US delegation demanded that developed nations be treated on an even playing field as the developed countries, which is like demanding that innocent bystanders are thrown into jail along with the criminals. China had other ideas: it demanded exemption from the treaty until it reached the level of development that the other rich nations had achieved – a decision driven by the fact it had an almost endless supply of coal hidden in its vast national borders.

The targets set at Kyoto were fairly simple: reduce emissions of six greenhouse gases – carbon dioxide, methane, nitrous oxide, sulphur hexafluoride, HFCs and PFCs. Collectively, this meant the Rich North had to reduce its greenhouse gas emissions by 5.2 per cent by 2010, compared with the 1990 levels. Norway, Iceland and Australia managed to beat the rap and actually win increases in their emission targets. The US managed to agree to a seven percent cut over the period but, considering it contributes around a third of all world CO_2 emissions, it is not difficult to see why many countries felt extremely hard done by the Americans. Still, the Kyoto Protocol was delivered: the rich countries were committed to certain and defined targets, more stringent than those given at Rio in 1992, whilst the poor nations were 'temporarily exempt'.

American conservatives once again stood up and rejected the steps that were being taken. They began a multi-million dollar advertising campaign, and lobbied against the Kyoto Protocol in government. Their warnings that environmental regulation would dramatically slow American manufacturing gained a foothold amongst the public, who particularly feared an increased tax on gasoline. After signing up to Kyoto, President Clinton lost his nerve in the face of a Republican-dominated Congress, and the Protocol failed to be ratified. The fact that the US failed to make any policy changes that would help to meet its targets gave other nations the excuse to fall behind too.

Still, the regular IPCC reports helped keep climate change at the forefront of the political and social culture, whether people would care much about it or not.

Then, after the most controversial of US Presidential races is history - probably the most controversial election victory in anyone's history - Al Gore lost to Republican nominee George W. Bush Jr. A new regime of conservatism took the hot seat at the White House, and the 'green-eyed' Gore no longer had any force at the highest level.

Bush's administration renounced the Kyoto protocols, saying that not enough evidence existed for global warming and that the steps to cut CO_2 would be too damaging to American industry. As the rest of the Rich North began to get moving on Kyoto and environmental regulation designed to combat climate change (not including Australia, which also rejected the Kyoto deal), the United States preoccupied itself with a war on terror – insisting that climate change was not caused by humans and therefore there was nothing we could do to stop it.

To this day, the actions taken by the world's nations are barely noticeable on a global scale, and the absence of an American part has caused any carbon-dioxide reducing targets to fall flat on their faces. Even if the willing nations of Europe cut greenhouse gas emissions to 1990 levels by 2010, there would be little point if the world's two biggest producers of greenhouse gases, the US and China are not doing likewise. Encouragingly however, in early 2004, David King, Chief Scientific Adviser to the British government proclaimed that the threat of rapid climate change was greater than that of terrorism to the wellbeing of the developed world. In a world completely gripped by the issue of terrorism, that statement means a great deal.

---

[i] Pearce, F. (2006). *The Last Generation*. Transworld Publishers, London. Pg 22

[ii] Retallack, S. (1999). 'How US Politics is Letting the World Down', *the Ecologist*, March/April.

Chapter 9 – Earth is Changing, Now

Recorded Temperature Rises

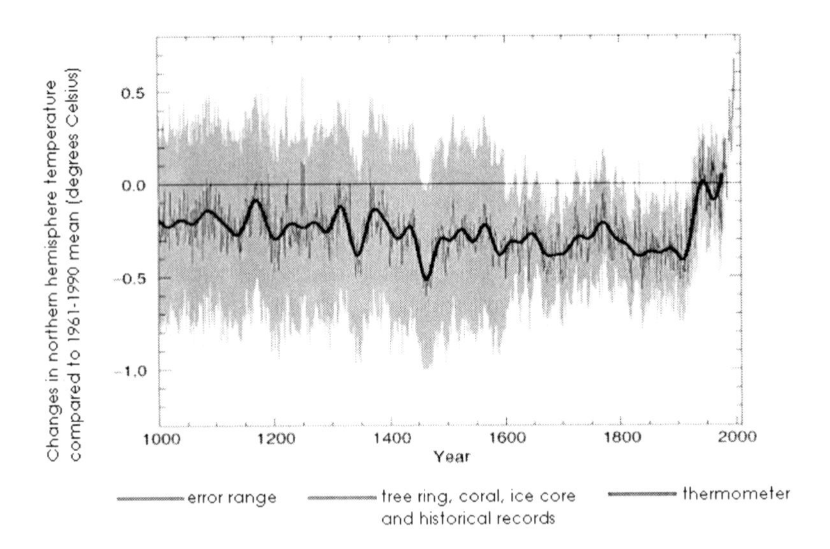

Figure 9.1 – The IPCC's famous 'Hockey Stick' graph of the last 1000 years. This data is derived from historical records, thermometers, tree rings, corals and ice cores.

For their 2001 assessment the IPCC published what became known as the 'hockey stick' graph of global temperatures, taken from a variety of sources. Because the data combined historical records, thermometers, coral, ice core and tree ring records it provides a very strong image. However, this image apparently proved too strong for some climate sceptics, who accused the IPCC of exaggerating the data to show an upturn in temperatures.

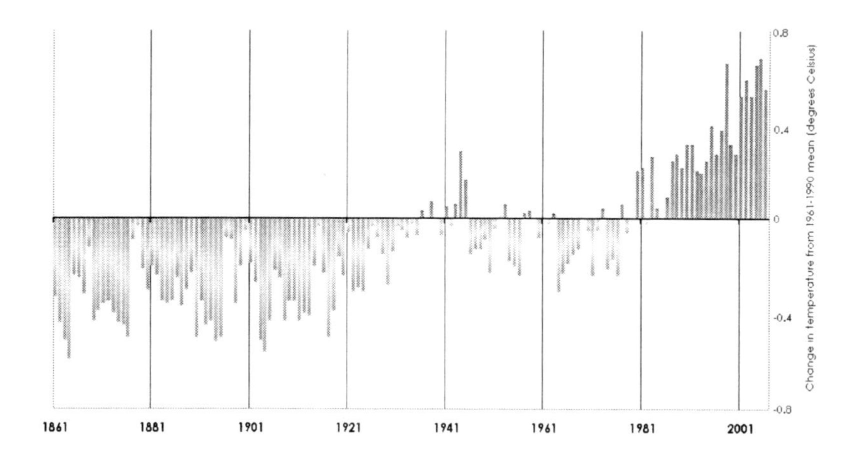

Figure 9.2 – The combined air, land and sea surface temperatures of the globe since 1861 show us global warming. There can be no denying, the planet has been warming since at least 1900. What many critics of global warming call a cold period, through the 1960s and 1970s, was actually something of a transition period.

If we focus on the last 150 years of global temperature we can understand better the most recent trends; the substantial rise in global temperatures since the 1960s contrasts with those of the period 1860 to 1910. But of the top 26 hottest years on record, 24 have occurred in the last 27 years, the ten hottest years on record occurred in the last 11 years, and (not including the freakishly hot year of 1998) the last six years have been the hottest six of them all. Each year the experts have been left more and more baffled; the temperature change, though it may fall in any given year, seems to be relentless in its desire to reach new heights. We could, before the decade is done, see a new all time record for global temperature and it is almost certain we will see that record broken before 2025.

As the 20th Century bade us farewell, it left with a 0.5°C jump in average global temperature over its final 25 years. In 2006, the UK Met Office saw the hottest July and hottest September ever in all of their history of recorded temperatures. The first half of the year was also the warmest on average than at any other time in the United States. Temperatures for the first six months of 2006 were averaged at 1.9 degrees Celsius warmer than the average for the twentieth century. This warming also came with below average rainfall, causing drought in over 45 per cent of the contiguous United States, some of it serious. However, in the north east of the country there was above average rainfall followed by severe flooding. It was also estimated that the drier conditions resulted in more than 50,000 wildfires during the first half of 2006.

In the UK, all chitchat eventually found its way around to the surprisingly mild winter that was taking place in December 2006 - February 2007 where hardly any snowfall was recorded for the entire winter. The New Year was celebrated with sunbathing up and down the east coast United States and parts of the UK and Europe as temperatures refused to drop. There were even reports of nights in early January being warmer than nights in the previous July, though we should be careful not to point to such short term anomalies as ultimate proof of global warming.

We should note that the Earth naturally swings from ice age to non-ice age thanks to the way CO_2 is concentrated and moved around. More importantly, a CO_2 level of 230-300 is optimum for our way of life because it provides temperatures that are not too cold and not too hot, promoting agriculture and giving us more land on which to live (avoiding high sea levels and monster ice caps).

Today, we have broken through this 300ppm barrier. In fact, we are fast approaching the unprecedented 400ppm point, entering un-charted territory for our species. The observed change can be seen as early as 1850 when atmospheric CO_2 concentrations had grown to 285ppm, crossing 300ppm at the dawn of the twentieth century. At the dawn of the new millennium, we are observing readings for 380ppm and more - straying a long way from what we know as our Holocene comfort zone.

If we put all this into perspective we can see the full effect of the change. During the last ice age there were about 400,000 million tonnes of CO_2 in the atmosphere – a figure that rose to 600,000 million tonnes when the ice age ended, the extra 200,000 million tonnes coming from the ocean. Since the Industrial Revolution humans have released a staggering 200,000 million tonnes extra, pushing the overall atmospheric total to around 800,000 million. In effect, we've added the equivalent of 25,000 million double-decker buses (or 29,000 million adult male African elephants) of CO_2 to the atmosphere, onto a pile that was only 75,000 million buses in the first place. Increasing any finely balanced mechanism by a third is dangerous. If you had a company and your costs had increased by 30 per cent over a short period you'd be more than worried. We humans have been quite lucky that the tranquil Holocene has lasted as long as it has – the last thing we need to do is start throwing things into its mix.

Wet & Windy

The word 'warm' is not a very helpful one when it comes to climate science. To any individual, 'warm' conjures up images of a crackling fire in wintertime, a comfortable blanket at night, a nice cup of cocoa, and so on. It's the same word we use to describe someone if we like them - if we think they are kind, friendly and generous. In science, the word 'warm' is avoided simply because it is too vague, inaccurate, and there are much better alternatives. However, few find a simpler and more understandable adjective to use when they compare what the world is like today to what it used to be like a few decades ago.

So what are the consequences of warmer temperatures? For some parts of the world they are much more drastic than at others. We can see large and obvious physical symbols of warmer temperatures with the melting of land glaciers. In places like Patagonia at the southern tip of South America, or atop Mount Kilimanjaro, this presents a big enough symbol of global warming. But in the Himalayas it becomes a massive human problem. Melt-water from the Himalayan mountain range supplies seven of the world's biggest and most important rivers, including the Ganges, the Indus, the Mekong and the Yangtze. These rivers then flow through densely populated countries around South East Asia, such as China and India, and provide more than half of the drinking water for about 2500 million people - that's about 40 per cent of the world's population. Without these glaciers in the Himalayas feeding these mighty rivers, those people could face a serious lack of fresh water within the next few decades.

Another example: the world's oceans are particularly partial to change if their temperatures rise only half a degree or two. Hurricanes, cyclones and typhoons are all the same climate phenomenon; depending on what ocean they originate from. The common term for the phenomenon in general is a 'tropical storm' and some may be too weak to classify as hurricane[1], cyclone or typhoon. Tropical storms that cannot sustain surface winds of more than 17 metres per second (39mph) are called depressions, but once it exceeds this speed, and keeps it up for a while, it becomes a tropical storm and given a name for easy reference. For a hurricane, cyclone or typhoon the storm must exceed wind speeds of 33 metres per second (74mph).

The name 'tropical storm' serves a descriptive purpose. All originate in the tropics, and rotate around a central 'eye' which can grow to several kilometres across. Though the winds are deadly around the rest of the hurricane, the eye itself is extremely calm – some of the lowest barometric pressures ever recorded have been from inside hurricanes. Most Atlantic hurricanes are born off the west coast of Africa, as thunderstorms, and move out to sea over the tropical waters. The link between the formation of storms and climate changes is not fully understood, especially when it comes to predicting whether a storm will transform into a hurricane. But scientists can draw a definite relationship between hurricane growth and ocean temperatures.

In the early stages of a storm, humid air from the ocean surface rises and its water content condenses to form droplets of rain and clouds. This condensation releases heat causing the cool air above to warm, and therefore rise. This rising air is replaced by more humid air from the ocean surface. This cycle repeats itself over and over, eventually moving heat from the ocean to the atmosphere and feeding the clouds into a storm. Without going into too much science, let us just say that as the surface temperatures of the ocean increase, the storm gains power. The more warm water it passes over, the stronger it becomes, which is when scientists get to the point where they have to start classifying hurricanes on a separate scale of intensity.

When Hurricane Dennis launched into Haiti and Cuba in July 2005, it caused widespread flooding and breakdown of basic infrastructure. The hurricanes that followed continued to throw themselves inland until all that was left, by the time the hurricane season eventually ended, were a lot of broken houses and a few million blank faces staring at the mess. But the summer of 2005 wasn't the beginning of the nasty turn in the weather - that point was marked a year earlier, in different places across the Earth.

---

[1] According to the US National Hurricane Centre the word 'hurricane' comes from the name 'Hurican', the Caribbean god of evil.

In 2004, Japan experienced a record number of typhoons, ten in total – more than any other year in recorded history. The scientific community was shocked when – for the first time ever, and going completely against what was ever thought physically possible – a hurricane appeared in the Southern Atlantic Ocean. Hurricane Catarina stunned Brazilians and took thousands by surprise and off guard. But it didn't stop there; 2004 also saw the all-time record for tornadoes in the US smashed to pieces.

The early part of the 2005 season saw the likes of Dennis, Emily and Irene cause chaos in the Gulf of Mexico, ripping up trees, sinking fishing boats, flooding coastal communities, blowing homes to pieces, cutting off electricity and water supplies, and even blowing huge, hefty oil rigs over on their sides and away from their safe grounding. Then, on 31st July, a study produced by Massachusetts Institute of Technology (MIT) reinforced the theory that warmer oceans, caused by climate change, are creating stronger and more destructive hurricanes. It found that major storms in both the Pacific Ocean and the Atlantic Ocean had increased in intensity, and the duration of hurricanes had gone up by about 50 per cent since 1970.

Then the big one. On August 25th 2005, just weeks after the MIT report was published, a hurricane formed off the East coast of Florida. When it barged its way through the Florida Keys and southern tip of the mainland, the storm caused billions of dollars' worth of damage and killed a dozen people, yet it was only big enough to be classified as a category one hurricane. The storm raced westwards until it cleared the land and found itself in the middle of the Gulf of Mexico. These waters were unusually warm, giving the storm more energy, more intensity, and more power. It began drifting north and went from a category 1 to a category 2, to a category 3... all the way up to a category 5 hurricane. On August 29th, the massive hurricane ended its two day journey over sea and landed just east of one of the biggest urban areas in the southern United States of America. This was Hurricane Katrina.

Beautiful waterfront houses were converted into piles of wood and wiring, roads were converted into rivers, fields into swamps, and vehicles into scrap metal. Water was everywhere, food and fresh water supplies suddenly became scarce, and shelter became the most sought-after of all. Thousands of people became homeless overnight, most having nowhere else to go. Millions of people hit the road, hoping to flee the initial storm, and then flee its devastating aftermath. Badly needed medical attention just couldn't be found for a lot of victims, and there were reports of families having to leave their lost loved ones at the side of roads because they had to keep moving to stay alive.

As thousands converged on the last bastion of descent shelter in the city - the Louisiana Superdome - in search of aid and food, as well as a place to sleep, the full scale of the tragedy came to light. An entire city, a huge metropolitan area, was under water like some modern-day Atlantis. It was strange to see in the age of technology and affluence that such devastation could still be wrought by nature, especially in the richest nation on the planet.

The story of Hurricane Katrina reads like some disaster in the Poor South. On Sunday 28th August a Louisiana newspaper, the Lafayette Times, reported that some forecasters had predicted that the levees outside the city may break under the force of the oncoming hurricane. Max Mayfield, National Hurricane Centre Director, tried to warn President Bush and Michael Brown (FEMA chief) about Katrina's force. A later media report picked up on this: 'On Saturday night, Mayfield was so worried about Hurricane Katrina that he called the governors of Louisiana and Mississippi and the mayor of New Orleans. On Sunday, he even talked about the force of Katrina during a video conference call to President Bush at his ranch in Crawford, Texas. "I just wanted to be able to go to sleep that night knowing that I did all I could do," Mayfield said.'[i]

Water was already toppling the levees on Sunday night, about the time 30,000 evacuees were gathering in the Superdome with about thirty-six hours' worth of food. At 7am the next morning, Katrina made landfall. A few hours later, despite everyone in the country hearing about the hurricane on television, President Bush attended a birthday cake photo opportunity with Senator John McCain. Then, as waters started breaking through the levees, Bush attended Medicare drug benefits in Arizona and California. At 8pm Louisiana Governor Blanco appealed for more help from the federal government: "Mr. President, we need your help. We need everything you've got."[ii] Instead of acting, the President went to bed.

On the Tuesday, the Pentagon claimed there were enough National Guard troops in the region, but later on mass looting broke out. The USS Bataan, a 844-foot ship designed to dispatch marines in amphibious assaults, had helicopters, hospital beds, doctors, food and could make up to 100,000 gallons a day of its own water, yet it remained idle just offshore of the disaster zone. In the afternoon, President Bush decided to attend another photo opportunity, this time strumming a guitar with Country singer, Mark Willis.

By Wednesday 31st August, the conditions in the Superdome were deteriorating rapidly. The Los Angeles Times reported: 'A two-year-old girl slept in a pool of urine. Crack vials littered a restroom. Blood stained the walls next to vending machines smashed by teenagers. "We pee on the floor. We are like animals," said Taffany Smith, 25, as she cradled her three-week-old son... At least two people, including a child, have been raped. At least three people have died, including one man who jumped 50 feet to his death, saying he had nothing left to live for. There is no sanitation. The stench is overwhelming.'[iii]

Bush - quickly realising the scale of the devastation - claimed that no one had expected the levees to break, despite his recorded video conference call to Max Mayfield. He then blamed the state and local officials for the troubles, stating that Governor Blanco never called a state of emergency. Over the next year, Bush would request a total of $62 billion for emergency relief, aid and levee repairs, and eventually he admitted that mistakes had been made at all levels of government.

Hurricane Katrina revealed two important things about our global society. The first is that we are simply not prepared for climate change, and certainly cannot cope with the devastation that the climate can cause. If one of the most affluent nations in all of human history can do nothing but sit back and watch as the winds and flood waters breach its defences, then there is certainly nowhere on Earth that has a viable chance of holding Mother Nature back. We've spent so long denying climate change, and then so long debating how to deal with it in future, that the future has already arrived, and we are caught with our guard down.

The second and perhaps most significant point is this: the victims of climate change are the poor. The very nature of being wealthy grants a person the ability to a) provide for themselves and their family when resources become scarce, and b) pick and choose where you live, either in a more solid home or in a completely different part of the country. Meanwhile the poor have no place to go and no means of keeping safe. How can a family without a car flee an incoming hurricane? Catch four or five buses out of town? What if the public transport is put on hold?

What do you do for food, medicine and shelter when you get away? Wait in line with a million other people for days, even weeks on end? Will that feed and clothe the children in the aftermath? Supposing you make it out of harms way to safety, and manage to feed and shelter your family for a few days whilst the weather calms down, what are you supposed to do back at home? How can a poverty stricken family afford a new home, or to rebuild their flattened or flooded homes, without money? What if your job no longer exists? How long can a person live off government help?

This is what happened to the people of New Orleans in the summer of 2005. They had to answer all of the above questions, and a lot more, or face starvation, homelessness and sickness. What if climate change turns nasty in Africa, or Asia, or South America, where poverty is rife and money is rare? What questions will the poorest people of the world then have to ask themselves? More importantly, will they have any questions at all, or will they just have to accept their doom? This is perhaps one of the most upsetting things about the whole climate change issue. The very people who made greenhouse warming happen will be the safest of all when climate change starts to bite hard. The first to fall will be the poor, the people who had virtually no part to play in creating the monster. A greater injustice I cannot think of.

Katrina didn't mark the end of the hurricane season. Three weeks later came Rita, also a category five hurricane, but one which, luckily, managed to hit the US in a less populated area. A few weeks later, Hurricane Wilma moved from Mexico to Florida, becoming the most intense hurricane ever measured! Only Cyclone Monica, measured off the coast of Australia in spring 2006, has beaten Wilma in intensity.

But Wilma didn't mark the end of the record-breaking 2005 hurricane season either! In fact, the end of the official 'season' didn't mark the end of the season, and hurricanes continued to crop up well into December. The WMO name hurricanes after letters of the English alphabet, starting with 'A', then 'B' and so on, until they get to the end. So the first of the 2005 season was called Arlene, the second Bret, the third Cindy etc. Missing out the letters 'Q', 'U', 'X', 'Y' and 'Z' this gives you 21 names to play around with -

usually more than enough to name all the hurricanes and tropical storms of the season. However, in 2005 there were 27 hurricanes and tropical storms, and for the first time the WMO ran out of letters. The final six were named after letters of the Greek alphabet: Alpha, Beta, Delta, Epsilon, Gamma and Zeta. With Cyclone Monica and the score of category five cyclones to strike Australia in the same year, 2005 turned out to be the year that oceans all over the world went a little crazy.

Some scientists contest the theory that warmer sea surface temperatures stimulate tropical storms. After all, argues William Gray of Colorado State University, it isn't just the sea surface temperature that causes a tropical storm to turn violent but the *difference* between the temperature of the sea's surface and the temperature at the top of the storm. If this is true, then warming of both sea and atmosphere will only keep the difference between the two the same – therefore not affecting the ability of a storm to turn into a hurricane, cyclone or a typhoon one jot.

But this is also contested. Many experts argue that the data supporting this idea is a little dodgy, or that the sea surface temperatures may well rise at a faster rate than atmosphere temperatures, therefore reducing the difference between the temperature at the bottom and the temperature at the top of the storm. Like many things in climate science this is a relatively new idea, and the chance of the argument being resolved soon is quite distant.

But the facts speak for themselves. Between 1995 and 1998 the Atlantic experienced more tropical storms than ever before over such a short period – a record forcefully blown away by the period 2004 to 2005. The year 1998 saw four hurricanes bounding around the North Atlantic all at the same time, an event unique for at least a hundred years.

Modellers at the Hadley Centre in the UK talked about the climate change-tropical storm relationship, and pinpointed a zone of ocean that could be the birthplace of hurricanes in the South Atlantic in a warmer world. Then, a few years later, Hurricane Catarina hit Brazil, forming very close to this zone. Unfortunately the Hadley Centre hadn't predicted anything like that to happen until around 2070! Once again, climate change had broken the laws almost as soon as we ever thought it was possible to do so.

Hurricanes, cyclones and typhoons are the epitome of nature's wrath. They are the physical manifestation of pure energy and force, and can be the most devastating of all weather phenomena. But a little further down the scale, into the smaller and less impressive displays of strength, we are still seeing unusual changes in the way the weather is behaving. A typical run-of-the-mill storm sucks up more moisture over warmer waters. As a result, storms are becoming heavier and wetter, and are more likely to drop all their moisture in one big go than in spurts here and there, as is usually seen. We are seeing heavier and prolonged rainfalls than ever before, because the sky is just too heavy to hold all that water up there.

This has caused more instances of flooding decade after decade, with increases seen on every continent. In some places, the seasonal transition from summer to winter has been disrupted, and, instead of white Christmases, people are seeing wetter and wetter ones. Snow forms when moisture trapped in clouds is frozen in cold pockets of air in the lower atmosphere; there are more cold air pockets in winter so there is more chance of snow. If clouds are too moist the water will fall before it has frozen, giving you rain instead of snow. Likewise, if there are less and less pockets of cold air around then there is less chance of snow.

Because of its frozen nature, snow settles on the ground and stays at least for a few hours, possibly for a few weeks or months depending on where you are. It gradually trickles away as the seasons warm up. Rain doesn't hang around; instead it falls and immediately tries to find its way back to the sea. As a result, less snowfall and more rainfall increases the number of flood events, because there is more moving water than static. In Asia, Europe and the Americas, the number of major flood events in the 1990s were at least triple what they were in the 1960s. The increase in flood events can be seen over every continent, and is consistently rising decade after decade, at no point falling.

In 2005, the climate continued to punish modern civilisation with a string of different catastrophes than those seen in the Atlantic and Pacific. Mainland Europe suddenly became inundated with floodwater, as did Asia. An unbelievable 37 inches of rainwater fell on Mumbai, India, in July 2005 in a single day. Water levels reached seven feet high and the death toll topped 1,000. Even in India, the hotbed of monsoon downpours, this level of flooding was unthinkable.

Great Oceans of Dust and Global Dimming

In summer 2005, the Shangdong and Sichuan provinces of China experienced huge and devastating floods. At the same time the nearby Anhui province continued to suffer from a drought that has afflicted the area for many years now. Widespread flooding and drought can occur in the same region simultaneously because this is how global warming works. It is easiest to think of human induced climate change as an exaggerator, making wetter places wetter, drier places drier, colder places colder and hotter places hotter. In fact, in some cases this is not true and places that were once permafrost, for example, are now becoming warmer. But there are places where this exaggeration *is* taking place, with the Sahara desert visibly seen expanding year after year, and flooding and monsoons in Bangladesh and India becoming much more frequent. So as global precipitation levels have increased by 20 per cent over the last century, some areas are actually getting less precipitation, and they are usually the places that were fairly dry to begin with.

As we have seen from the evidence of storms and flooding, higher atmospheric temperatures increase precipitation around the world. The evidence of droughts, also worldwide, shows that higher atmospheric temperatures are also causing moisture to relocate and concentrate itself away from some areas. Desertification is now occurring across all the major continents, with particularly harsh or persistent drought afflicting the Western Sahel, Sudan, Ethiopia, Southern Russia and Chile.

In the Poor South, farmers are under pressure from economic markets to produce more and more agricultural products year on year. Pushing land beyond its nutrient capacity is not uncommon, and may understandably compound the drought problem. We can also see evidence of droughts in rich nations, where precipitation has decreased. Places like Italy, Spain and Portugal, Southern Japan and parts of the Eastern United States are witnessing the economic costs of less precipitation.

Historically of course, droughts have always been a part of human history. We only need to look back at the Anasazi of the Colorado Plateau, the Mayan people of the Yucatán Peninsula, or the great cities of Mycenae and ancient Petra, to see that our species has always been affected and influenced by a lack of rainfall. Since 1800, droughts and famines worldwide can usually be attributed to a human cause, be it mismanagement of land, clearing forests leading to the removal of underground water tables or overpopulation. Droughts that have occurred for the last few decades, and those that will occur in future, will only be aided by these factors, not created by them.

The last time the world took notice of a widespread drought was during the 1980s, when North East Africa, just along the transition zone between the Sahara to the north and the vegetation zones to the south, became a scene of human catastrophe. This famine was not caused only by climate changes - more a mish-mash of terrible factors that all came together to kill a lot of people. Still, the single biggest reason for the drought, and subsequent famine, was a persistent and devastating lack of rainwater.

The region around Ethiopia and the Eastern Sahel is accustomed to drought, and infrequent years of low crop yields. It hasn't always been this way, with evidence suggesting that droughts in this part of the Sahel being a modern phenomenon rather than historic. Here, the vegetation frontier creeps up and down the continent in cycles, shrinking the desert from May to September and enlarging it during the rest of the year – like a wave on a beach.

The annual movement of the vegetation frontier is powered by the summer monsoon trough, which shifts northwards during summer in the northern hemisphere when the waters of the North Atlantic warm. The region near the equator, where northern and southern trade winds meet, is called the intertropical convergence zone (ITCZ). The ITCZ moves north and south as the seasons pass, trying to keep in line with the region receiving the most solar radiation. Using a clever mechanism involving moisture, pressure and temperature, the ITCZ brings a late summer rain to the Sahelian region, providing human populations as well as vegetation with some much-needed fresh water. During the early 1980s, the ITCZ consistently failed to move as far north as usual, leaving parts of the Sahel dry and desperate.

In 1981, a hard drought hit the Eastern Sahel, and the following years were almost as poor. The rains of 1984 failed on a much bigger scale, plunging the entire region into famine, as farmers didn't have enough water to grow their crops or feed their livestock. In Sidamo, a major area of food production in Ethiopia, disease ravished the crops. By March, the spring had been so dry for Ethiopia that the government declared that only 6.2 million tonnes of grain could be produced, leaving around five million Ethiopians facing starvation.

Eventually the madness passed, with around a million people dead from starvation and a further 50 million afflicted in some way or another. Other large-scale famines to strike the Horn of Africa, including Ethiopia, occurred in 1987-88, 1991-92, 1993-94, 1999 and 2002. What's more, one of the most terrible consequences of the famine of 1984, and those since, is that a lot of people in the Sahel are almost completely dependent on foreign food aid - meaning that they are putting their life in the hands of someone else to provide surplus grain. If a large drought hit the US or Europe, or anywhere usually accustomed to providing aid to Africa, there would be less surplus grain and therefore less aid.

However, in the Rich North we often blame the famines of the early 1980s on overpopulation and bad farming methods, leaving us safe in the knowledge that Africans had been the authors of their own doom. In reality, the blame lies squarely on the shoulders of the Rich North, and this takes us back to the Global Dimming phenomenon we touched upon at the end of Chapter 6.

Scientists had been following a progressive trend for years during the late 1980s and early 1990s, showing that the amount of sunlight reaching Earth's surface had decreased substantially over the twentieth century. This seemed to suggest that the intensity of the sun was falling and, if so, then the whole idea of global warming would be thrown into doubt, since such a large reduction in sunlight would have overridden most artificial warming committed by humans. However, during the mid 1990s, Professor Veerabhadran Ramanathan, one of the world's leading climate scientists, had noticed a substantial drop in sunlight over the northern islands of the Maldives, which got their prevailing winds from India in the north. Meanwhile the southern Maldives, which got prevailing winds from the Antarctic way down in the south, had experienced virtually no sunlight drop. It soon became obvious: the thick pollution spewing out of Indian power stations was acting as a veil over the northern islands, blocking out much of the sun.

Pollution that caused this global dimming effect comes from particles of sulphur dioxide, soot and ash that collect in the atmosphere and increase the amount of sunlight that is reflected back into space. The fossil fuels used in India caused the Maldives to darken by ten per cent, but what about the fossil fuel pollution from North America and Europe? By applying Professor Ramanathan's findings to his own research, Dr Leon Rotstayn found the answer. The particle pollution from these heavily industrialised parts of the world were changing clouds over the North Atlantic and cooling the ocean. The ITCZ was no longer attracted to the north and dodged away southwards, preventing the vital rains from reaching the Sahel. In short, the heavy addiction to fossil fuels in the Rich North had directly caused the devastating droughts and famines that plagued the early 1980s and filled our news reports with pictures of skeletal babies and weeping mothers. It was our fault.

But you only have to look at the consistent droughts of the 1980s, 1990s and beyond 2000 to realise that Africa is not suffering a transient problem. In fact, the word drought suggests some sort of natural rhythm, with an eventual return to normal after a short while. The exceptional run of droughts in northern Africa since the 1970s indicates that the problem is not merely transient, but rather a large-scale permanent shift in climate has taken place.

The outlook for Africa today is ever bleaker. Malnutrition on the continent can be widespread, even in years when production appears to meet the needs of the people. According to the UN's Food and Agricultural Organisation (FAO), between 40 and 50 per cent of the sub-Saharan population goes hungry every year, and the region is 'worse off nutritionally today than it was 30 years ago.' Sub-Saharan Africa has the largest undernourished population than anywhere in the world. Though UN figures show a worldwide average fall of malnutrition from 37 per cent to 18 per cent over the last three decades, sub-Saharan Africa has failed to do better than its 1969 figure of 34 per cent.

Of course, all this malnutrition and famine isn't completely a global warming issue. Largely it is a human issue, caused by poor farming practices as much as poverty, corruption and greed. But as the twenty-first century comes into full swing and climate change begins to bite harder, the forecast is less and less precipitation, leading to more and more drought. Any drought in this region may trigger drought on a massive scale.

After drought comes famine, and violence usually follows. The Darfur region of Sudan experienced devastating droughts leading up the explosion of violence that tore across the region during 2004 and 2005; no doubt the severe lack of food and water had compounded the civil strife and in-fighting leading to civil war. In Malawi the rains failed during spring 2005, leaving food production dangerously low, and would have sentenced more than five million Malawians to death if alternative solutions had not been found. To the north of the Sahara desert, the situation is not much brighter. Morocco and Tunisia have experienced around a 30 per cent decrease in precipitation over the last 100 years, and, together with Libya, the total agricultural land lost in this region to desertification is around a quarter of a million acres every year. Pile on top of all these problems the stresses facing wildlife and you have yourself a real problem, especially in a continent where tourism is the major economic driver and where the natural wonders of the land are by far the biggest tourist draws. If ecosystems begin to collapse, as they have been doing in recent years, the most famous of the money-spinning species in Africa - such as elephants, lions and gorilla - could be lost forever. With further desertification of arable land, and drying of lakes and streams, the entire African continent could be plunged into economic and humanitarian chaos.

A well-documented case of desertification in Africa is the disappearance of Lake Chad, on the borders of Cameroon, Chad, Niger and Nigeria in central North Africa. When full, Lake Chad was the sixth largest lake in the entire world; it proudly wore this honour up until the beginning of the 1960s. Today the lake is barely more than a very big puddle, surrounded almost completely by dusty sands and dry vegetation. So how has the decline of the sixth largest lake in the world to nothing more than a large body of water impacted the region, particularly since it occurred over such a short period of time?

Before the 1960s Lake Chad was the lifeblood of the region, providing water for drinking, irrigation, livestock and a source of fish. Millions of people surrounding the lake, in all four countries, depended on its existence for their livelihoods, either directly or indirectly. In Niger, the city of N'guigmi was once surrounded by the waters of Lake Chad on three sides, but today over 60 miles of desert lie between it and the Lake.[iv] What greater symbol of climate change is there than a city surrounded by water in the first part of the century and being surrounded by sands in the second half?

Traditional fishing grounds are now gone, leaving fishing boats and fishermen stranded. The human dangers of climate change have really begun to show in some places too, like when the fishermen of Nigeria followed the receding waters over their border and into Cameroon. Cameroon was becoming desperate itself, and didn't appreciate the Nigerian fishermen intruding on their territory and harvesting fish stocks that they badly needed themselves. The result was sporadic fighting, as well as the fallout on an international diplomatic front. In places where farmers have begun to cultivate land that used to be the lake bottom, all sorts of issues relating to property ownership and rights have exploded into the foray. The overall situation is becoming increasingly tense.

The winter of 2005-2006 saw Somalis suffer the effects of a relentless lack of rainfall, forcing many to trek up to 70km (44 miles) to find water in temperatures as high as 40°C. "Many families are surviving on a small jerrycan of water for three days," said an Oxfam spokesman. "Our assessment team gathered reports of people being forced to drink their own urine because of their desperate thirst."[v] This was the worst drought to hit Somalia in forty years, and was the unfortunate overspill of a drought begun in late 2005.

Desertification has been increasing decade upon decade for at least the last thirty years as a global average. It happens when warmer atmospheric temperatures suck moisture out of the land, as it does the oceans. Instead of black, moist soil, full of nutrients and water, what you get is dry clumps of mud and cracked land, where no plant can take root. In the 1970s, around 625 square miles of land was becoming desert every year; in the 1980s, the figure was almost 850 square miles. In the 1990s, the average annual area of land that had turned into desert was 1375 square miles, an increase of 36 per cent on the decade previous, and a 120 per cent increase on the 1970s rate. Desertification is happening more and more, but more importantly, it is happening faster and faster as time goes by.

In their October 1999 impact assessment, the UK Met Office's Hadley Centre have modelled the effect of climate change on water supply over the coming century, in three scenarios: 'business as usual' where nobody does anything about CO_2 emissions, a doubling of pre-industrial CO_2 levels at 550ppm, and a tripling of pre-industrial levels at 750ppm. In both the 'business as usual' model and the 750ppm model, the number of people facing a persistent shortage of water will be 3,000 million, about half the world's population today. Under the 550ppm scenario the number falls to a no more comforting 1,000 million people.

For wheat crops, a higher atmospheric CO_2 level is a good thing; for everything else it is bad. With more carbon dioxide the photosynthesis process is speeded up in cereals such as wheat. This boosts plant growth in temperate regions, because they are able to make more food for themselves and grow larger and faster. The fossil fuel industries are quick to leap on this fact to promote an image of a utopian future where there is enough food for everyone, all thanks to their multi-trillion dollar industries and the pollution they cause. But, in truth, the benefits to production of such crops will only be very small, and when compared with the other effects of climate change in future – sea level rise, desertification, etc. – they are virtually worthless. Furthermore, not all food crops respond to more atmospheric CO_2 in such a way; maize, millet, sorghum and sugarcane are just some that are actually hindered. At the same time, many aggressive weeds do benefit, and then there is the issue of crop-eating pests that will potentially flourish under warmer temperatures.

Following the drought there is the added risk of fertile topsoil being washed away during the next rains. This can devastate farmland almost as much as the lack of rainwater itself. As farmers wait patiently for rains to come and water their land, the soil dries up. If it gets very dry, soil becomes very light and crumbly, and at the first sign of rain it may be washed away into rivers and streams. This is the same problem that occurs on land that was previously forested but has been cleared. The lack of roots holding the soil together will allow the soil to be washed away quite easily given enough rain. Soil erosion like this can also, in some cases, lead to landslides, and particularly in areas of the planet that experience El Nino. Landslides can obliterate towns and villages lying in the way. Towns around the Alps will be more susceptible to landslides through floods and precipitation, such as Italy where a landslide around the turn of the millennium reached 100m deep.

The 1998 Hadley Centre assessment predicted 30 million additional people going hungry by 2050, and a further 18 per cent of Africans facing the problem.[vi] But all this is just a global average, and global average does not account for the niggling little details. Some areas would feel the bite harder than others, and during the severe drought or flood events the problem could leave the 'increased average' trend line and head straight for the disastrous. Fundamentally, agriculture doesn't just rely on how much rain falls, rather on when and for how long. Climate change could devastate localised regions that have always been stable and dependable places for growing crops, even in the birthplace of civilisation and agriculture itself, the Fertile Crescent.

The issue of desertification and drought is core to every human alive. At first only the poor may be affected, but as climate change keeps growing, farmers in Europe, Northern Asia and North America will certainly see their yields decline. Farmers have it hard enough as it is today, with government quotas and market prices dictating their every move – imagine how many farmers are going to make ends meet when fields begin to dry up on them. Aside from the economics and the rich-poor issue, desertification affects us all because it damages our ability to provide our most basic need in life: food.

At the Poles

If you've ever seen an image of the world from a climate model in a newspaper or on television - with coloured blotches indicating the amount of warming that will take place and where - you will have noticed that the redder (hotter) places are at the north and south poles. As time moves more deeply into the future, the red becomes stronger and more intense, while many other parts of the world may be orange or even yellow still. This doesn't mean that the north and south poles are going to become warmer than the tropics, but that they are going to experience greater temperature rises than the rest of the world.

On the face of it, this all seems very lucky for us humans, since very few people actually live in and around the Polar Regions compared with the rest of the world. However, the fact that the Polar Regions are going to warm more than any other over the coming century is of absolutely no benefit to life on Earth whatsoever. In fact, more warming at the poles will *take* time *away* from us rather than grant us extra, and it is really the worst place you could pick for temperatures to soar. This is because the poles are especially sensitive to even the smallest temperature changes. Around the fringes of the Polar Regions the melting has been going on for many years now. No other place on Earth is so sensitive to such a small change in mean temperature, and few are so melodramatic about it.

If one were to see photographs taken at the Arctic (at the North Pole) and the Antarctic (at the South Pole) it would be virtually impossible to tell them apart. Both are white, cold and very, very big; in terms of appearance the two could quite easily be twins. Under the surface, however, and the two are completely different – more like brother and sister. The Antarctic is actually a huge continent of solid land surrounded by ocean; the Arctic is a huge body of frozen ocean surrounded by land. The ice covering the Antarctic land mass is around 10,000 feet in depth, while the ice at the Arctic averages out at less than ten feet deep. This incredibly slender layer of ice at the Arctic makes it extremely vulnerable to changing temperatures, and throughout history it has always been this ice to the north that has truly been iconic of climate changes, whether it engulfs the land around it, including northern Europe, or whether it disappears completely.

The first successful journey to the Pole is credited to Anglo-American Navy engineer Robert Edwin Peary, who claimed to have reached the Pole on April 6th 1909, accompanied by African-American Matthew Henson and four Inuit men named Ootah, Egigingwah, Seeglo and Ooqueah. This was quite an accomplishment for the times, especially considering the gear they had and the speed at which they traversed the last few days, but Peary picked a good time for it. Januray temperatures at the Pole can range from -43°C to -25°C and this can make adventures next to impossible. In summer, temperatures average about freezing point (0°C) but this means there may be more meltwater pools on the ice further south, or unstable ice fragments, which can add many days to your journey if you want to get around them. In March, when Peary set off, temperatures are much more favourable.

According to the Worldwatch Institute report of March 2000, the ice sheet that covers the Arctic Ocean has lost 40 per cent of its volume since the 1970s and could be completely gone within a matter of decades. Though this wouldn't affect sea level (the ice is already floating) it will signal the beginning of the really big climate changes that are going to affect this planet.

Surrounding the polar ice, on the land, the air is so cold that the moisture in the ground actually freezes, creating what scientists call permafrost. Though it makes for a harder living, permafrost doesn't make life in these parts completely impossible, for plants or humans. Large areas of land surrounding the Arctic Circle are frozen for the majority of the year, but since permafrost occurs in places such as Sweden, Finland, Canada, Alaska and Siberia, it is safe to say that human civilisation has managed to find a way to survive quite successfully with permafrost about. In fact, humans have been living in these areas for many thousands of years, and not once has permafrost caused a mass exodus.

Humans have become so greatly adapted to living with permafrost that its removal today, and over the past few years, by warmer temperatures, is dramatically threatening life in some places. When the climate warms in these cold parts of the world, the permafrost inevitably begins to melt. What you get is mud, and plenty of it, and a muddy region is a much less desirable place to live.

Buildings that are built especially for permafrost ground are slowly sinking into the new mud, or collapsing piece by piece; they were built with foundations specifically to suit a hard ground, regular soil just won't do. In some places, whole sections of land are sliding away as the permafrost turns to soil: buildings sitting above are left to slide away also, usually in fragments. Houses, apartment buildings and office buildings are starting to be abandoned in parts of Alaska and Siberia; places like airports and hospitals are feeling the pinch. Roads are thawing all over permafrost regions, creating problems for industries that depend on trucking for importing and exporting. There are few better ways to travel in frozen areas than by road, so the issue of mudslides and water-logging is a very real one which such communities have to take seriously.

Nature is also showing battle scars in its fight against a rapidly changing climate. Some tall tree species are especially adapted to pushing their roots deep into frozen soil, clamping a firm anchor to grow tall and strong. The thawing is now starting to make their firm grip in the soil less firm, and many are beginning to lean, and even topple over, as the years pass. As the new generations of seeds fall into the ground they will find it impossible to grow successfully in the soils, and as time goes by the forest frontiers will begin to retreat to smaller and smaller areas of suitable land. Soon there will be whole areas of bare land that were once great forests, thousands if not millions of years old. In other parts, the seasons are changing fast. Spring arrives much sooner, and autumn comes much later, extending the warm period of the year, and thawing more permafrost with each passing orbit around the sun. Ecosystems – including human communities – will have to adapt very rapidly to the changes, or else face collapse.

As we have already seen, the value of ice sheets to the workings of the climate system is invaluable. Ice is white, and as such, ice sheets act as gigantic mirrors, reflecting most of the solar radiation that hits it and sending it back out into space. This is why you often need sunglasses in places that are covered in ice – if the light (and heat) were not reflected then the ice would be black instead of white.

To recap, the reflection of radiation by ice sheets creates a positive feedback loop. If the temperatures are colder, then the ice sheets will grow; larger ice sheets reflect more radiation, which then makes temperatures colder. During periods of cold global climate, the presence of large ice sheets can further encourage the low temperatures. Equally, during periods of warm global climate, the absence of these huge mirrors can lead to more absorption of solar radiation and warm the planet more. This is because both water and land absorb more radiation than ice.

If the polar ice sheets were to melt slightly then there would be more radiation absorbed by the water surface. As the waters then begin to warm, it triggers more melting of the ice sheets, exposing more water, absorbing more radiation, and melting more ice sheets. This runaway loop is vital to any scenario where the global temperatures are seen to rise, because it spells the beginning of an accelerated warming process. It is especially vital in the Arctic because the ice sheets are so thin and more sensitive, which means that the acceleration of rising temperatures would be triggered quite easily. And guess what... this runaway loop has already begun in the Arctic, kicking off a heap of trouble for anyone living south of the North Pole.

And that brings us nicely to the polar bear – perhaps the most famous inhabitant of the Arctic and one of the first big casualties of climate change, now and in future. For many, the demise of the polar bear is an inevitable consequence of global warming, and when I say demise, what I really mean is extinction. The World Wildlife Fund (WWF) established its Arctic Programme only a few years ago in an effort to follow and track polar bears and their cubs. To do this, they have to fire anaesthetic darts from a hovering helicopter and then quickly attach thick collars around the necks of the animal that then gives off a homing signal. At the same time, whilst the great bear is still unconscious, they make measurements of its limbs, head and torso, estimate its weight and assess it for any illness or wounds that may not be obvious to them from the air. And then they run like hell, for a polar bear in a daze is still a polar bear capable of killing a man with one swipe.

By attaching a homing signal to the polar bears, the scientists can track their movements throughout the year, and assess how human disruption is affecting their natural world. By taking measurements of individual bears the scientists can assess if any of them suffer from a lack of food and if the numbers of malnourished bears is on the increase. Unfortunately, the WWF only has enough funding for twenty flying days every year to make these measurements and collar the bears.

Polar bears are not complicated creatures, although they are completely outstanding. Their main source of food are seals, which they reach by bashing holes in the ice to the waters below and waiting for an unsuspecting seal to appear at the surface for air – effectively conning their prey to fall into their pounce. In order to bash through the ice to create a hole to the sea below, the ice has to be thin enough, as well as – obviously – being over sea and not land. However, since the 1990s an area of sea ice as big as Sudan has been lost to temperature rises. Under the most likely scenarios for climate models, by 2100 the entire area of sea ice in the Arctic could be gone during summer months (it retreats slightly every summer but never actually disappears) ultimately condemning the polar bear to extinction. However, it is worth remembering a very recent NASA-funded report that suggested that the ice sheets atop Greenland could vanish every summer by as early as 2040 – leaving us with the tiniest of windows in which we can slam on the brakes and stop this from happening.[vii] If we don't, the polar bear will be nothing more than a fond memory.

This is a dramatic change, not only because it transforms vast areas of the planet into something completely different, but also because it is going to happen over such a rapid timescale. Indigenous communities of the Arctic have already begun noticing these changes in temperature and sea ice. Ice seems to be breaking up earlier in the spring, winters seem to be warmer than previously known, and ice is staying thinner all year round on a whole. The link between temperature and sea ice is best made when you consider that air temperatures have increased in the Arctic by 5°C on average in the last 100 years and that arctic sea ice has decreased in volume by 14 per cent since the 1970s.

Estimates show there are around 20,000 to 25,000 polar bears left in the wild at present, 60 per cent of which are in Canada. In the Hudson Bay region of Canada, sea ice is now melting earlier in the spring and forming later in the autumn. As a result, the bears have less time on the ice for which they can hunt and store their energy in their fat for the lean summer months. Already scientists are seeing polar bears becoming thinner thanks to an inadequate food supply, and infant mortality is on the rise.[viii]

When the ice breaks up the bears have to make a dash for the nearest bit of land. Sometimes the ice is so thin between the land and the edge of the ice that it breaks up before they turn back, and they may face a swim of several miles to reach home. For every week the ice breaks up earlier in the Hudson Bay, the bears are seen to come ashore around 10kg lighter.[ix] Less time, less food, less offspring: eventual extinction.

And just to rub snow in the face of the polar bear, we humans also contribute to its downfall in more direct ways. Industrial pollution from North America, Europe and Asia drifts northwards and is breathed in by the polar bears, which are not used to breathing anything but clean, crisp air. We hunt them too; and here comes the killer punch: we also extract oil from the Arctic, adding to the already overwhelming odds stacked against the polar bear. Onshore oil developments exist in Alaska, Canada and Russia. At present there is a single offshore oil site in the Arctic, in the Alaskan Beaufort Sea, but exploration has taken place in the Kara, Pechora and Barents Seas, the Sea of Okhotsk and the Davis Strait in the Canadian high Arctic Islands. Further development can be expected in future.

Oil from spills reduces the insulating effect of bear fur when the fur becomes covered. To compensate, the bears are forced to increase the amount of food they have and so boost their energy. With little access to food anyway, this becomes a serious problem. Ingested oil can cause liver and kidney damage, and it is a common enough occurrence, especially when there are several ways in which a polar bear can ingest oil: grooming its fur or eating contaminated prey. Then you've got the excess pollution from the oil extraction process which produce toxic substances, and the everyday operation of an oil facility such as traffic, construction, seismic blasting and so on; all of which help bring down the greatest predator of the Arctic.

The Polar Regions are beautiful – anyone can see – and especially the Antarctic. Where else can you be surrounded by bright sunlight, freezing cold temperatures, howling winds, sharp bottomless crevices, and immense white landscapes that stretch off into the distance like some Star Wars-type ice planet, floating through space in some galaxy far, far away? Where else can you be stood surrounded by frozen water and still say that you are in a desert? Where else is completely dominated by H_20 yet receives less than two centimetres of precipitation per year?

Explorer Ann Bancroft once said of the Antarctic: "You're going uphill, chasing the horizon. Sometimes it's above your head, at your midsection, or beneath your feet, but you never catch it... The wind swirls particles of ice and snow, and the sun catches them and you see reds and turquoises and purples. Each day is remarkable in and of itself." As Roald Amundsen remarked, "the land looks like a fairytale." Or perhaps not, as Ernest Shackleton famously said: "No person who has not spent a period of his life in those stark and sullen solitudes that sentinel the Pole will understand fully what trees and flowers, sun-flecked turf and running streams mean to the soul of a man."

If the Arctic is the fragile sister of the two Polar Regions, then the Antarctica is the big bad brother. Though colonies of penguins, seals and birds are found on the fringes of the frozen continent, the middle is completely devoid of life. The gangs of scientists that inhabit the ice mass come from more than a dozen countries, but if you want to take a tour to the pole itself (yes, there really is a stripy barber pole) then you have to be prepared for altitude sickness. Few people actually realise it, but the ice cap is so thick in the middle of the continent (around 10,000 feet) that one would need the same physical preparation for a journey here as one would to climb Mount Etna in Sicily. In fact, the average altitude on Antarctica is much higher than on any other continent. So much snow has fallen here over the last 50 million years that it has pushed the surface higher and higher into the sky, and forced the bedrock underneath further and further down with all the weight. The Antarctic bedrock is now largely below sea level because of the enormous weight of the ice above it. Can you imagine how much force it takes to push an entire continent downwards, even just slightly?

There are three physical features to take note of in Antarctica. The first is the long mountain chain that stretches down the middle of the region, from

Victoria Land to Coats Land. These Transantarctic Mountains boast the continents two highest peaks – Mt. Markham (4351 metres high) and Mt. Kirkpatrick (4528m). The second feature is the Ross Ice Shelf, an area of permanent ice that – unlike the majority of the ice sheet – hangs over ocean water. It is significant because it pushes far inland until it only about 500km from the South Pole itself, and is virtually flat in the shadow of the huge mountains to its west.

The third feature that is physically unusual is the Antarctic Peninsular, a sharp tusk of land jutting northwards towards the southern tip of South America. Only a fraction of the peninsular is actually ice that sits atop land surface – the majority is made of patches of ice shelves. The biggest shelf is the Larsen Ice Shelf, which expands and retracts naturally with the changes of the seasons. So far, the Larsen Ice Shelf has remained more or less intact over the years, but to its immediate north a smaller ice shelf – named Larsen-B – is gradually shrinking as time goes by. Larsen-B has decreased by around 60-70 per cent since 1995. This is no mean feat, as the Larsen-B ice shelf contained so much water that it sat around 200 metres above the surface of the seas surrounding it. The energy it takes to melt or break this ice is nothing to be sniffed at.

Scientists guessed that Larsen-B would still be around by the year 2100, with or without global warming, but they had gravely misunderstood the nature of a melting ice sheet. The photo below shows the Larsen-B ice shelf in January 2002 with its white surface dotted by what appear to be holes leading straight down to the ocean below. In actual fact, those black spots are pools of water that have collected on the surface of the ice. Ice doesn't just melt from its fringes, but it also melts from the top downwards. Most of the time, natural melting pools sink back into the ice and freeze, whilst only some keep sinking until they create a hollow tube through the ice. Under warmer temperatures more and more of the pools sink without freezing, creating more holes in the ice. Eventually the entire ice sheet is incredibly unstable, and will inevitably break up and drift away into the sea. This is what happened to Larsen-B in February 2002, over the course of just four weeks. Millions of tonnes of fragmented ice spilled into the Weddell Sea, drifting away until they, inevitably, melted completely. It was as if someone had pulled the bottom card out of a house of cards and the whole lot caved in on itself.

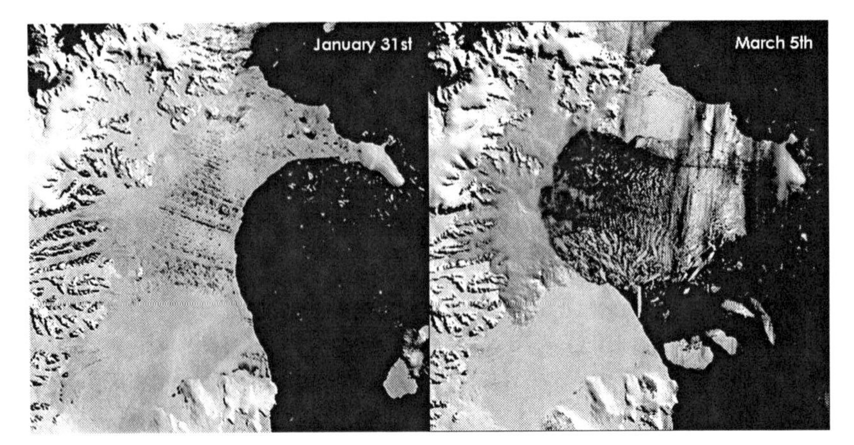

Figure 9.3 – The crumbling of the Larsen-B ice shelf, Antarctica. The entire sea ice shelf disintegrated in just over four weeks. (NASA photos).

The damage caused by ice sheets melting on this scale is great. The sudden influx of fresh water into the ocean disrupts the ocean conveyor. For the tiny[2] Larsen-B ice shelf, the implications of the melting are probably not significant – although it may be a few years before we truly know how big the disruption has been. However, if similar melting happens in the Arctic, and continues to happen around the Antarctic, then the consequences will be a lot higher.

Fortunately, as anyone who has sat and watched ice melt in their drink will know, if ice floats on liquid and then melts, the water line will not rise or fall to any significant degree. Ice shelves exert nearly all their weight on the ocean below, pushing the sea level up slightly, so when they come to melt, the sea level does not rise any higher. For land ice the story is different. Because it is supported by land – either at the top of a mountain range or sat squarely over the land like the Antarctic and Greenland – land ice sheets do not exert any influence over sea levels when they are stable. When they melt, however, water that has previously been outside the ocean is suddenly added to it - increasing the total volume and raising the sea level.

Antarctica can be divided into two zones, East and West, along a theoretical line between the inner-most points of the Ross Ice Shelf and the Ronne Ice Shelf. The East Antarctic zone is the largest mass of ice anywhere

[2] Tiny from one point of view – Larsen B at full strength was, after all, bigger than Northern Ireland.

in the world. Recent studies have shown that it is beginning to *lose* mass around the fringes, and that the total volume of ice is on the decline. More ominous is the state of the West Antarctic sheet, which is rather less sturdy than the other zone. Here the ice shelf is bedded in against a set of smaller landmasses – like islands – as well as part of the 'mainland' that the East Antarctic sheet sits on. Because it is held in this way by land, it exerts little pressure on the surrounding sea, as an ice shelf would. But because it is propped up against these smaller landmasses it becomes more fragile, and if it happened to slip off and fall into the sea, scientists estimate it would raise sea levels worldwide by around six metres.

In Greenland, the same problem of melted pools of water that occurred on the Larsen-B ice shelf is now also seen to be occurring. The effect is identical. Huge rivers of melt-water are forming on the surface of Greenland, sometimes creating lakes before they push on down to create hollow tubes in the ice sheet - what scientists call moulins. With enough crevasses and moulins, lot of water gathers beneath the ice sheet, between itself and the landmass below. After a while, there is so much liquid water between the ice sheet and the landmass that the ice sheet begins to slip, breaking off in huge sections that slide away towards the sea over many years. Today we live knowing that more land ice is melting than we have ever seen before, indicating the beginnings of a rapid turn of events that could lead to the long-term loss of over half the Greenland ice sheet.

At this point, let us stop and take stock of the situation: our species is causing the ice sheets at both poles to melt away, fragment and crumble. This is cause for astonishment on its own, regardless of the consequences. It is truly shocking that these glorious geographical features, millions of years old, are being transformed by our species... without us even laying a finger on them. We are actually changing the Earth here, and there is no greater symbol of humanity's destruction of the planet than the unprecedented rate of collapsing ice sheets on both sides of the world. The ice sheets are the canaries in the coalmine for the global climate, and by melting away they show us that climate change is not a problem for tomorrow, but a problem for yesterday that is happening today. The canaries have stopped singing.

Nature and Disease

Of course, one of the many problems created by changing temperatures is that the natural arrival of each season is also changed. All over the world the climate is being pulled and stretched into new distortions, and all this is playing havoc with the changing of the seasons.

The changing of the seasons affects all ecosystems. When seasons change there isn't a sudden and rapid transformation from one type of weather to another. Instead, you get a slow transition from warm to cold, or vice versa. Plants are wonderfully adapted to sensing climatic changes; flowering plants begin to bloom and deciduous trees produce their first leaves of the year only when the temperatures are consistently high enough. In autumn, when temperatures start to fall again, trees will shed their leaves, creating wonderful, fiery colours as the leaves shrivel and fall. But this fall only occurs if temperatures are consistently low enough, and a particularly mild winter can see leaves still clinging to trees even as spring starts to come around.

Virtually everything in the natural world gets its cue either from the rising and falling temperatures, increasing or decreasing precipitation, or from other creatures that have already begun to act. The changing of Earth's seasons is like climate change but on a miniature scale, and since the tiny wobble of the planet that causes seasonal variation has been going on for all of its existence, life on earth is completely dependent on it. Birth and rebirth of millions of species simply could not happen without the annual to-ing and fro-ing of the warm and cold months.

Climate change is beginning to shake the transition between one season and another, smudging the lines. If the climate stays warm for longer at the end of summer, the transition into autumn and eventually winter is delayed. Similarly, spring arrives earlier as warming brings about a faster retreat of winter. In some places it is the other way around, with shorter summers, later springs, earlier autumns and longer winters.

Ecosystems are incredibly delicate. If one species is adapting better to the shifts than another, that second species could be in for quite a lot of trouble. In his book, 'An Inconvenient Truth', Al Gore highlights the case of migratory birds in the Netherlands.[x] A quarter of a century ago, birds that had migrated south during winter were arriving back in the Netherlands around the 25th April. Six weeks later their chicks would hatch from their nested eggs, the peak number hatching on 3rd June, just in time for the peak appearance of caterpillars, on which the chicks feed. Today, the warmer temperatures have caused the peak caterpillar date to move forward to 15th May. Since the migrating birds take their cue to fly home from weather thousands of miles away they are still arriving around 25th April, leaving their offspring with a little over two weeks to hatch and catch the caterpillars. Of course, they cannot hatch faster than is natural, so although the peak hatching date has moved forward to around 25th May, they are still rather late to catch the caterpillars. Mother birds are finding it impossible to find enough food for their offspring, and as a result many chicks are dying from starvation.

Some reptilian species have a strange adaptation to temperature that might single them out for a quick decline in a warmer world. The eggs of the American alligator produce females only when they are hatched at temperatures lower than 32°C, otherwise it's a boy every time.[xi] This same problem seems to occur again and again in some reptile or aquatic species.

The case of the golden toad of Costa Rica's Monteverde Cloud Forest Preserve is thought to be the first documented victim of extinction purely thanks to climate change. The cloud forests depend on humid conditions to sustain the local ecosystems, and rising temperatures are being blamed for causing humidity levels to fall since the 1970s - pushing many species beyond recovery. But there have been more victims in the Monteverde Preserve alone, including twenty of the area's fifty species of frog and many forest lizards, whilst several species of birds (including the keel-billed toucan) have begun to intrude on other's territory, forcing both original occupant and the invader into a difficult struggle for limited resources.

Climate change seems to be killing off species in various different ways, but most of the time species are being pushed to extinction indirectly. When temperatures rise and the Earth's zones shift pole-ward, many species begin to move to keep up with the shift. This creates invaders and interlopers, which then force an original species out of an area. Usually this occurs with insects, birds and plants. Species that are bound by physical limitations of land geography or national park fencing cannot really go anywhere to keep up with the changes, become trapped, and are beginning to decline in numbers in the face of their invasive contemporaries. Others are just trapped in the face of a new habitat type.

It is thought that the Arctic Tundra, a breeding ground for many bird species, could start relinquishing land to northward moving forests. Since the Arctic tundra is at the top of the world, and there is nowhere else to go, WWF estimates that around 40-60 per cent of tundra could be lost, meaning around half of all water bird populations could disappear within the next few decades. The species at most risk are those already threatened – the Emperor Goose, the Tundra Bean Goose, the Red Breasted Goose, the Greenland White Fronted Goose and the Spoon-Billed Sandpiper.[xii]

Few species manage to keep up with climate shifts, and extinction occurs more often as a result of those species losing their source of food. If a certain plant is forced out of an area then some species will struggle for food unless they can follow it. This indirect form of extinction places the most dependent organisms at the top of the risk list, and the first victims of climate change *en masse* could be the mammals for this reason.

Now, of course, life on Earth has been migrating here and there since time immemorial, in response to the naturally-changing Earth. Gradual changes in the environment allow for gradual changes in species to adapt. However, there have been huge and catastrophic upheavals in Earth's past that have resulted in massive extinction events or 'Era changes'. Humankind has inadvertently set the ball rolling for a new and cataclysmic extinction event, and, like a rolling snowball, it keeps getting bigger the longer we allow it to go on.

The evidence for this is coming in all the time from people studying species in the field. Because climate is changing rapidly compared with its natural variation, most species are unable to keep up with the rate of adaptation both biologically and in terms of migration. A WWF report from 2000 pointed out that plant species would be required to migrate ten times faster due to human-induced climate change than they had to during the last ice age.[xiii] Fossil records indicate that the fastest plant species on the planet are capable of two kilometres of migration a year, whilst the slowest may only be capable of about 40 metres; when you consider that temperature changes may force plants to shift at between 1.5 - 5.5 kilometres annually, the issue becomes all too clear. This spells the end of the road for millions of species if climate change continues, and the worst-case scenario envisages another mass extinction and inevitable Era change.

There are other problems that don't originate from species starving. In some places, the changing of the seasons wards off pests, which helps protect areas of forest and vegetation from long-term destruction. Cold weather in particular is a primary reducer of pests that would otherwise run amok as their populations swarm. Warmer winters are allowing this to happen, and in some areas there are signs that damage to plant populations by invasive pest species will take a long period to overcome, and that's if the pests stop staying longer, which they won't do as long as temperatures suit them.

Every year the world uses millions of tonnes of pesticides, yet more than 40 per cent of the world's food production is lost to the pests, weeds and plant diseases. This costs the global economy more than $500 thousand million every year. As temperatures keep on getting higher, and winters become increasingly milder, the planet will become more suited to food pests. They will be able to move to higher latitudes, and higher *altitudes*, for longer periods during the year. Where insect larvae are usually killed off by cold in winter, milder winters will allow more to survive, ready to begin a major infestation at the start of the new cropping season. A rise of just 1°C would see the European corn-borer chew its way 500 kilometres further north than usual, and great airborne armies of locusts could be a regular thing all across the northern Mediterranean, devastating thousands of square kilometres of crops.

As we know, creepy crawlies such as mosquitoes and ticks can carry infectious diseases, many of which are extremely harmful to human health. Some such diseases are well known: E.coli, Lyme Disease, Influenza and Tuberculosis, while there are many others that are less publicised but ever as dangerous. Human beings are at least risk to disease during the cold months, or in colder climates, because disease carrying species generally prefer warmth (mosquitos, bats, and tsetse flies are some). Though there are some exceptions (rodents, lice, fleas and snails are rather less picky about their temperature) it all comes down to how well the viruses and germs that they carry fare in varying temperatures. Global warming has changed all this.

With it becoming warmer at higher latitudes, the carriers are spreading away from the equator into their expanded dominion. Similarly, warmer temperatures also allow carriers to inhabit higher *altitudes*, pushing viruses and diseases out of the flat lands and river valleys and into the hills and mountains. The capital city of Kenya, Nairobi, sits above the natural mosquito line, propping its inhabitants high above the sea of infection. Since the 1980s, however, the mosquito line has shifted higher, and today the city is staring disease squarely in the face. Finally, on top of all this, the carriers have the opportunity to hang around longer than usual as the higher temperatures stall the winter season, giving them more time to wreak havoc.

The last century has seen the highest number of species extinctions since the retreat of the last ice age. The rate of extinction on the planet has been rising since the start of the Holocene, but it is only since industrialisation (around 1800) and the growth of cities, population, industrial scale agriculture and manufacturing, that the 'background rate' (the natural rate of extinctions, those that occur whether humans are around or not) of extinction has really been broken. You only have to look at any of the big extinctions (such as the dodo) or near extinctions (white rhino, giant panda) of the last couple of centuries and you see humans are either hunting them to the edge, hunting their *prey* to the edge, or destroying the habitat until there is nowhere for them left to live. In short, humans have drained life from this Earth through everyday actions.

However, the increase in extinctions we saw in the last 200 years won't amount to much compared with what we are expecting to see in the next 100 and beyond. In fact, we are looking at a future where the rate of extinction, per year, is going to rise from around 2000 species in the year 2000, to about 50,000 in the year 2100. The reason is climate change. Climate change pushes the boundaries of a normal, predictable climate out of the physical boundaries of an ecosystem. Species have to adapt quickly to changing circumstances, and for some the demand for change is so quick that they simply cannot cope.

For example, life in the African savannah is already quite tough. For a family of elephants, survival depends on their ability to walk hundreds of miles in the dry winter months, searching for the old familiar water holes where they can quench their thirst. The grass and flood plains of spring and summertime are long gone - dry dusty seas squatting on the edge of what is essentially the desert. The migration to water holes in colder months is already precarious and risky, with many individuals dying along the way, getting lost, or being hunted down on arrival by desperate lions who are also trying to use the water hole. A warmer climate is pushing the extremes to their limit. The savannah grass is drying sooner, and the number of useful water spots is shrinking. Elephants are starting to travel their regular winter journey only to find that there is no water there for them to drink. In the African savannah water is a vital commodity, and a severe lack of it is beginning to thin out many populations.

Savannah regions are now slowly becoming too arid for their natural inhabitants and, likewise, many arid regions are becoming too deserted for species adapted to arid regions. In the rainforests, a changing climate is beginning to drive changes in places far beyond the reach of human influence. More CO_2 in the atmosphere is fertilizing the fast-growing species of plant that zoom upwards and crowd out the slower-growing ones. Biodiversity is diminishing, as many sources of food are slowly pushed away. Increasing CO_2 in the air may seem to be a good thing for plants in general, but many researchers are now seeing plants growing weaker and poorly formed. The plants may be getting all the CO_2 they need, but if there is no equivalent increase in other nutrients then they cannot be as nutritious to herbivores that depend on them. It's like adding more sugar to a child's daily diet – they may still have the energy to grow and move, but unless you increase other vital nutrients at the same time all you get is a sickly, unhealthy adult.

Another significant change that we are witnessing takes us back to the poles. The fall of the emperor penguin is of grave significance to the natural world. Since the 1970s, a warm spell has descended over Antarctica (warm for Antarctica anyway) and it has begun thinning many of the ice shelves that overhang the ocean. From a first instance this may not seem like much of a bad thing for the emperor penguin – after all, it is the only species of bird that can live and hunt completely in the ocean, without ever needing to rest on land. But the thinning ice is proving to be devastating because they need ice to breed and nest. As the ice becomes weaker it breaks apart, and can sometimes drift out to sea.

Emperor penguins are hardy creatures. They form loyal partnerships for life with their mates, and look after their offspring in every waking moment. In the early part of the year they march about 90km inland to reach their breeding site, where they begin courtship about April. As early as May or June the female will lay one egg, but then has to make a hasty retreat back to the sea to replenish her nutritional reserves. With incredible care, she will transfer the egg to the male, who will incubate it for more than two months without food, surviving off its own fat reserves and by spending most of its time asleep to conserve energy. The transfer of the egg is tricky – if the egg is out in the cold for more than a moment the chick will be lost.

The males then show a great amount of brotherhood by huddling together through the harsh, cold gales. They also have the responsibility of looking after the chick if it hatches before the mother returns, and they do this by sitting the chick on their feet, covering it with its pouch and feeding it. Ideally, the female will return with a full stomach and take over the care of the chick. Then it is the father's turn to go eat. His journey is shorter than before because the summer heat melts some of the ice between the breeding site and the sea. When he returns, both parents tend to the chick and then all three will return to the sea. In a matter of months the whole process will start over again – with all penguins over four years old going inland to breed.

Persistent warm temperatures since the 1970s have begun to ruin this epic tale of family unity and instinctive survival. Warmer temperatures and stronger winds thin the sea ice, sometimes breaking it off and sending the breeding site floating out to sea. Without stable breeding ground the penguin cannot breed. If the breeding ground is reduced in size then so is the penguin population. Sometimes the penguins, eggs and chicks may find themselves drifting out to sea as the ice thins and breaks up. The upshot of all this is that the population of the species has declined by around 70 per cent since the mid 1960s.

Nearly half the world's original forest cover has been lost in the last 50 years. The word 'lost' falsely gives the impression that no one is to blame – as if we just misplaced a few billion hectares of trees. Whilst forests burn or topple over, or are simply cut down, we have to remember that deforestation accounts for about a fifth of all greenhouse gas emissions, and – more importantly – our present forests take in between 20-25 per cent of all our CO_2 emissions. If these remaining trees are plunged into a dangerously new climate they fail to grow successfully, and start to release more carbon than they take in - this has already been recorded in places like Alaska.

The pressures of deforestation and habitat destruction are deadly when combined with the added force of climate change. In 2000, the percentage of species classified as threatened by the IUCN, was bigger than anyone had ever imagined: 12 per cent of all known birds, 21 per cent of all known amphibians, 24 per cent of all mammals, 25 per cent of all reptiles, 29 per cent of all invertebrates, 30 per cent of all fish, and 49 per cent of all plants were seen to be at risk of being lost.[xiv] Today those numbers are higher and still rising.

Like the tropical forests, coral reefs are tremendous 'hot spots' for biodiversity on the planet. They are like vast cities for marine life in an ocean that is largely empty. Thousands of different species might be found within a small patch of coral alone, and many are still undiscovered. As a child I always dreamed that I would one day travel to the Great Barrier Reef, off the coast of Australia, and swim with the brightly coloured fish and the strangely shaped creatures that slink around down there. I would see the wonderful underwater plants, and the organisms that look like plants but really aren't. Now it seems that little dream might be gone. The world's corals are dying and have been dying for quite some time now: in the Caribbean, the Red Sea, around Madagascar, around the islands of Oceania and Australia.

During 2005, the world's coral collection took a huge blow. This had been the hottest year on record and, at time of writing, the extent of the damage on the world's coral reefs was not really known. However, we do know that 1998 was the second hottest year on record, and that year saw a coral loss of over 16 per cent. Of course, like nearly everything in the environment, the destruction of the coral reefs can be pinned down to a number of causes: pollution, more acidic waters in some areas, and dynamite fishing in places where there is no law against such practices. However scientists are beginning to see that coral reefs are dying largely as a result of warmer ocean waters.

Most mental images of coral reefs conjure up all sorts of bright and diverse colours, weird and wonderful shapes, big and little fish, and snakey, pointy, tiny, flat or spiky little critters that hide in the coral and in the sand around. It is therefore very difficult to imagine a coral being almost completely a monotonous greyish white colour, the same tone as we associate with the dead. Since coral 'bleaching' was unheard of before the twentieth century, and extremely rare until the 1970s, we must allow ourselves to become very, very worried about the sudden epidemic we see today.

When corals whiten – a process known as coral 'bleaching' – it is like a body being stripped of its meat until all that is left is bone. Tiny organisms live in the transparent membrane that covers the coral 'skeleton' – giving coral its rich colours and indeed its life. These tiny organisms are extremely sensitive creatures, and become highly stressed out by too much heat. When they exodus in search of more suitable temperatures, all they leave behind is the milky skeleton. If the coral skeletons are left for many years without creatures living in them, then they will degrade, break up and disappear forever. That is why marine-biologists become very concerned when coral bleaches in the first place, and become more concerned when it continues to do so year after year.

But it doesn't stop there. A warmer ocean temperature is only one way in which corals are being killed by climate change. The second way is to do with the fact that the world's oceans are taking in roughly 30 per cent of all extra carbon dioxide added to the atmosphere by humans. When CO_2 dissolves in water it increases its acidity, and corals do not fare well in acidic waters. The lower pH affects the saturation levels of calcium carbonate in the ocean, which corals rely on. Many other organisms rely on oceanic calcium carbonate to make their shells or hard outer surfaces, and more acidic water is disrupting this process for them all.

As well as serving as vital break-waters, protecting many important areas of coast, coral reefs are more miracles of nature, they support the majority of life-forms in the oceans and any marine-biologist would tell you that they are the true biodiversity hotspots of the seas, and need protecting at all costs. The fact that they are dying is a crime.

The Great Barrier Reef, off the coast of Australia, is the most vulnerable reef to the warmer waters and increased acidity. In 1998, more than four-tenths of the reef here was bleached, and, in 2002, around six tenths had been affected in some way. Ocean waters, particularly in the south Pacific, are sensitive to the appearance of El Nino and both 1998 and 2002 were El Nino years. Public outcry at the damage to the reef forced the Australian government to take action and they quickly set about protecting around a third of the reef from commercial fishing and limiting tourism there. But it wasn't fishing or scuba-divers that were bleaching the coral, it was climate change. To save the coral, the Australian government should try to tackle the bigger problem but, as yet, Australia remains one of only two Rich Northern countries that consistently refuse to ratify the Kyoto Protocol, as well as the one of the biggest CO_2 emitters in the world per person.

Dr Terry Done, acclaimed coral reef expert, estimates that an increase in global temperature of just one degree Celsius would leave more than eight tenths of the world's reefs bleached – just three degrees would see the end of the bright and beautiful coral reefs we know today.[xv] When you consider the delay period in the climate system (roughly thirty years for ocean waters to catch up with atmospheric temperatures) we may have to admit that even if we slammed the brakes on hard today we might not be able to prevent a widespread dieback of reefs the world over.

The 0.5°C (0.9°F) jump in global temperatures, over the last 25 years of the 20th century, equates to a 2°C rise per century, that is, provided that greenhouse gas emissions don't accelerate, which they are. Though this 2°C figure stands below the predicted change seen in models (because it is only an extrapolation of a short term trend and therefore not likely to be accurate) we can still make one useful observation: it is estimated that most ecosystems can only manage a temperature change of 1°C over a century, so 2°C or more could place many ecosystems between a rock and a hard place.

At last, we must recognise that climate change threatens human populations as well as the rest of the natural world. This is, of course, something we already know, but who will it affect and how? Just to emphasise the fact that humans are part of the natural world too, let us look at one human case study: the Kalahari San.

The San survive by hunting and gathering roots and tubers. Some sources estimate that the San's culture goes back sixty thousand years and some say a hundred thousand, but, either way, theirs is definitely the oldest in the world. The San once lived in Southern Africa but were forced northwards into the Kalahari by Bantu tribes and white farmers who wanted their fertile land. The San are a migratory people, moving about with the wild game in bands of about thirteen individuals. They shelter in temporary housing made from a semi-circle of tied branches covered in turf. There are no leaders in these bands and hunting is a collaborative effort.

Today there are only a tiny proportion of San left in existence. The reasons are not so much environmental as political. The Kalahari basin consists largely of what is known as Botswana today, Southern Africa. At independence from the British, Botswana was one of the poorest countries yet, at present, thanks to the discovery of large diamond deposits, it is one of the fastest growing economies in the world. Unfortunately, prominent economists in the government failed to appreciate the value of the nomadic San, who lived off the land and never needed money or advanced technologies. Governments have forced the San to relocate to permanent locations, which goes against their nomadic nature, and for compensation they've been forced to join the rest of us too, using piped water, electricity and other modern amenities.

Now this may seem like a very charitable thing to do - providing health care and modern civilisation for a people lacking it all – but it's really very terrible. By forcing the San to join modern civilisation, they are no longer allowed to San people. They don't want running water, packaged food and electricity. Some don't even want the health care, preferring natural remedies they've discovered over the millennia. Trapping them in 'free housing' eliminates their hunter-gatherer livelihoods, because they have to keep moving to keep up with the change of nature.

Without leaders, the San find it difficult to express their grievances at this debacle. Without concept of land ownership, they find it impossible to prove that land is theirs, leaving their ancient homeland open to outside interference. The San population has fallen extremely low in the last century, because of absorption and annihilation; the tragedy is what could happen to the ancient San when global climate change swings into action.

Future climate change will likely dry Botswana and the Kalahari basin, causing the die-back of vegetation presently covering the stable dune systems. The destabilisation of the sand dunes would wreak havoc on the arable land within the basin, and turn it into an uninhabitable dustbowl. In the past (before around 1100 BCE) the San survived past climate swings by simply migrating to better areas; but present-day political boundaries will prevent them from doing this in future, and it is very likely an entire people and their way of life will be dispersed and lost, thanks to no fault of their own. That is the tragedy.

---

i Lush, T. August 30th, 2005. 'For forecasting chief, no joy in being right.'

http://www.sptimes.com/2005/08/30/State/For_forecasting_chief.shtml, Accessed 16th Apr 2007.

ii Thomas, E. (2005). 'How Bush Blew It: Bureaucratic timidity, bad phone lines, and a failure of imagination. Why the government was so slow to respond to catastrophe.' Newsweek. http://www.msnbc.msn.com/id/9287434/site/newsweek, Accessed 17th Apr 2007.

iii http://thinkprogress.org/katrina-timeline, Accessed 16th Apr 2007.

iv Gore, A. (2006). An Inconvenient Truth, Bloomsbury Publishing, London. Pg 117.

v Radnedge, A. 'Somalis dying as drought worsens,' Metro, Friday, Feb. 17th 2006.

vi 'Climate Change and its Impacts', the Hadley Centre for Climate Prediction and Research, the Meteorological Office, November 1998.

vii Science Daily webpage, http://www.sciencedaily.com/upi/index.php?feed=Science&article=UPI-1-20061212-12390200-bc-arctic-ice.xml, Accessed 16th Apr 2007.

viii Climate.org webpage, http://www.climate.org/topics/climate/polarbears/shtml, Accessed 10th Dec 2006.

ix World Wildlife Fund (WWF) Arctic Program: http://www.panda.org/about_wwf/where_we_work/arctic/polar_bear/threats/climate_change/index.cfm, Accessed 10th Dec 2006.

x Gore, A. (2006). An Inconvenient Truth, Bloomsbury Publishing, London. Pg 152.

xi Flannery, T. (2006). The Weather Makers. Penguin Books Ltd, London. Pg 90.

xii 'Climate Change Threatens Rare Arctic Water Birds,' World Wildlife Fund, 3rd April 2000, in: Godrej, D. (2001). The No-Nonsense Guide to Climate Change. New Internationalist Publications Ltd. Oxford. Pg 77.

xiii 'Global Warming and Terrestrial Biodiversity Decline', WWF, August 2000.

xiv IUCN World Conservation Union Red List: www.redlist.org/info/tables/table1.html, from Smith, D. (2003). The State of the World Atlas. Earthscan Publications, London.

xv Woodford, J. (2004). 'Great? Barrier Reef.' Australian Geographic 76:37-55.

Chapter 10 - Tomorrow's World

What will Earth be like next century, and next millennium?

Climate models suggest that to have a complete melting of the Greenland glacier, temperatures in the region would have to be at least 3°C warmer than they are today, though 7Ma (the last time the glacier was absent) they may have been about 5°C warmer. Taking into account that higher latitudes experience warming two or three times greater than the planet as a whole, temperatures 7Ma must have been around 2°C to 2.5°C warmer than today on a global average. Measurements of atmospheric CO_2 at this time indicate levels of around 500ppm to 600ppm.

Predicting the world's climate for the coming decade is not like predicting the weather in your local area for the coming week. It takes a lot of computing power to work out all the scenarios, of which there are millions. Over the years, computing power has grown, and climate modelling has become a lot more accurate - although simultaneously a lot more complicated.

One of the complications is a changing human world. The human population at the moment is in the middle of a tremendous boom. We've gone from a population of just 1,000 million in 1804 to a population of 6,000 million in 1999. Within less than two centuries we have seen another 5,000 million people explode onto the Earth, when it had taken a little over two million years to get the first 1,000 million.

More people means more mouths to feed, more homes to build and more livelihoods to create. Our world today is so lop-sided in terms of wealth distribution that we struggle to find enough of these things for over a third of the people alive today. When you consider that the population is forecast to hit around 9,000 million by 2050, and finally settle down at something like 9,500 million around 2060 it begs the question: *how on earth are we going to provide*?

We've got to provide enough farms and rice paddies, yet protect our forests and woodlands from being destroyed. We've got to provide jobs and housing for the people to avoid widespread poverty, yet we have to make sure our urban zones don't completely wipe away the countryside by over-expansion. We have to do all this and still hope that there are enough of today's natural wonders left for future generations to enjoy and learn from. Even without climate change these problems are enormous; *with* climate change it seems the solutions are impossible. But we can only know the answer if we take into account two vital variables: the most likely future we will face, and the likelihood that we as a people can change this outcome.

Predicting the future of an ever-changing world

There are enough fossil fuel reserves to last us another few hundred years, and unless we can drastically cut the amount of carbon emitted into the atmosphere, then atmospheric CO_2 levels will increase to possibly five times higher than those before industrialisation. At this volume, carbon dioxide levels would be comparable only to those last seen 100Ma in a 'greenhouse earth', where there were no polar ice sheets, higher sea levels and very different ecosystems. It would take roughly 1000 years for all this excess carbon to be removed from the atmosphere by the gradual intake of the oceans. Our world would be a very different place if we burnt *all* of the fossil fuels we have beneath our feet, and that different world would become the new norm for thousands of years. Then again, our present day climate may not return at all. After all, the Holocene is one of the most strange and unusually placid times in Earth's climatic history, so we could hardly expect it to find such favourable balance again in the near future.

It all really depends on greenhouse gases. At the dawning of the year 2000, atmospheric CO_2 levels were rising at a rate of 15ppm per year. Recent studies suggest that this rate has risen since then, but really the question is whether the rate will continue to rise into the future... and by how much. Since we cannot *know* this, we have to satisfy ourselves with making best guesses, and the only way to do that is to understand three key factors:

1. How efficiently we will burn fossil fuels in the future;
2. How much the human population increases in future;
3. How much carbon energy each person uses in the future.

The first, efficiency of fossil fuel usage, is one of the most interesting to look at. For a start, we can expect our use of oil and gas for energy to peak around

2020-30 at present rates of use. As reserves of oil and gas decline, and we use less of them, it is highly likely that coal will take over as the primary source of the world's energy. Since oil and gas are considerably cleaner fuels than coal, once again entering the Age of Coal would be considerably disastrous for our climate. Even if it fails to become the primary source of the world's energy, it will still stand as the primary source of the world's *carbon* energy, and prolonging, if not worsening, the problem of CO_2 in the atmosphere. This becomes especially critical when you consider that when our species first started burning coal it was the high-grade black (anthracite) type, which is far cleaner than what is left today - the low-grade brown (bituminous) type that produces far more CO_2 for the same amount of fuel.

Those of a technocentric mind will point to humanity's ability to solve any problem through innovation, and say that we will one day be able to burn fossil fuels with very little carbon emissions at all. There could be some process in power plants, or in the combustion engine, that captures the CO_2 and prevents it pouring into the atmosphere. On the other hand, such processes are yet to be invented, and exactly how long will it take to invent them? Perhaps too long.

Furthermore, there is no certainty whatsoever that Poor South countries will adopt new technologies in future as they demand more and more energy. Some may find it easier and cheaper to continue burning fossil fuels as they are, just as Rich Northern countries have been doing for a couple of hundred years now. It would therefore be quite foolish to rely on future technological advances to clear the atmosphere of more CO_2.

And the issue of Poor South countries leads us to the second factor: human population. By the time the population boom ends around 2075, we can expect to see human population reach between 9,000 million and 11,000 million - almost double what it was in 1990. On its own, in a world without fossil fuels, this would be a titanic problem, but when you consider that every one of those people will want to devour energy at faster and faster rates you really start to see the predicament we face.

This then takes us to our third factor: how much carbon each person will be responsible for in the future. In a world of expanding population we already expect to see more energy consumed, but we are also seeing material standard of living rise too. For some of the 'non-category' countries – those that cannot be classified as either Rich North or Poor South – the increase in material standard of living will push them more towards the Rich North bracket, and many will reach it within the next few decades. For the Poor South, a high material standard of living will be achieved probably around 2040 or beyond, and this is going to keep the rate of CO_2 emissions per person increasing at a higher and higher rate well into the long-term future. Ultimately, what we can expect to see is semi-industrial economies becoming fully industrial, and agricultural economies becoming semi-industrial, within the next 50 years. Only a few decades from now, agricultural economies will number less than ten, and that is when the bottom drops out; when industry takes over today's poor, when more and more people demand more and more things, and when more factories spring up with no certainty that they will choose clean energy over dirty energy. If you think we've got a CO_2 problem now, just wait until the Poor South stops being so poor, demands more televisions, more computers and more cars.

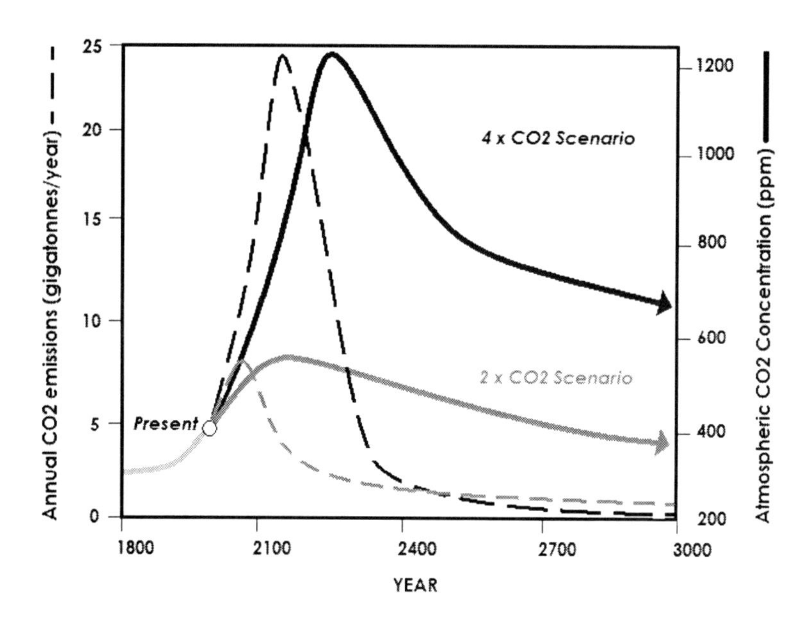

Figure 10.1 (above) – Two possible scenarios of future carbon dioxide emissions due to human sources. One looks at a doubling of pre-industrial levels, which would take a rapid and sustained international effort to achieve. The more realistic value is a quadrupling of pre-industrial levels, probably by the mid 2100s. Also plotted, (unbroken lines) are the subsequent atmospheric concentrations due to these emission scenarios. If concentrations quadruple, they will reach levels not seen on the planet for many millions of years.

There are two scenarios that stand out for the future of carbon emissions. The IPCC projects the range of CO_2 outputs leading to the year 2100, and other leading scientists have continued these projections well into the centuries to come. The two broken lines in Figure 10.1 illustrate two future scenarios with very different mindsets. The upper line shows what emissions would look like if humans continued to pump out CO_2 with only economic consideration at heart, and minimal efforts to curb fossil fuel usage. This line shows an increase of CO_2 emissions to around 25 gigatonnes per year by the latter part of the next century, before declining steadily over the following hundred years to pre-industrial levels. This scenario presumes we burn fossil fuels until we literally cannot burn them any longer.

The second line is more optimistic, and as some have argued, maybe *too* optimistic. It shows what will happen if we take action against our fossil fuel addiction, peaking around 2030 before falling again below today's eight gigatonnes per year to five or six gigatonnes per year by 2100. Though this line peaks so soon in our future, it really doesn't mean anything until we take the delay in climate response time into account. Emissions may peak by 2030, but how long will it take for the climate to catch up, and what will the effects be?

With carbon *emissions* we can expect to see the numbers increase or decrease in future with a little bit of fluidity; there will probably be a long-term change in fossil fuel use that has an effect on emissions rather than sudden and rapid changes, so the curves are likely to be quite smooth. With CO_2 *concentrations* the predictions become more complex because we do not know how the climate will respond to changes.

For the immediate future we can probably forecast that the atmosphere will continue to take in just over half of CO_2 emissions as it does today, with the rest being absorbed by the oceans and the biosphere. At this point, we also have to consider how much of our biosphere will remain in the future and at what rate deforestation will occur. We know that when oceans take in CO_2 their waters turn slightly acidic. Within a few hundred years, the peak CO_2 emissions will be mostly absorbed into the oceans, increasing their acidity. This will eventually lead to fewer organisms being able to survive in the oceans and again this will lead to a reduction in the amount of CO_2 the oceans take in – a feedback loop.

Figure 10.1 also charts the likely levels of atmospheric concentrations for each of these two scenarios. For the doubling of pre-industrial CO_2 levels (the $2xCO_2$ scenario) the optimistic of the two, we can see concentration values peaking around the year 2200. The other line shows concentrations reaching four times the pre-industrial level by 2300. Both decline slowly, though of course it takes longer for the quadrupling of CO_2 levels (the $4xCO_2$ scenario) to reach pre-industrial levels once again.

Comparing the lines for emissions and concentrations we see the significance: there is a delay between emissions and concentrations - meaning whatever we do today we pay for tomorrow. You can also see the time scale involved. It may seem pointless looking almost 1000 years into the future, but it shows just how big an impact our present-day world is having on the planet. Carbon dioxide concentrations are likely to be unnaturally high well beyond the year 3000 – that is about a big a footprint as you can leave.

Surprise!

There are a couple of little surprises that might warrant a mention. First, we should take note that global warming will not be a linear process. We know this already from our look at the geological past and the fits and bursts that seem to happen all over the place when it comes to global climate. There are several climate tipping-points - critical thresholds that, once crossed, will open up the floodgates and unleash a new wave of climate changes. A light switch needs a certain amount of pressure to actually switch – any less and it stays still. There are many climate switches all over the planet, ready to activate these tipping-points. At the moment, scientists believe we are not applying enough pressure to these climate switches, but as time goes on, and our legacy of greenhouse gases grows and grows, we are exerting more and more pressure until inevitably they must flick.

One of the possible switches may be the rainforests. At present, we like to think of our rainforests as being permanent absorbers of carbon dioxide, and our allies in the fight against climate change. Recent studies suggest that the largest rainforest, the Amazon, is not a stable and mature forest with growth and decay in balance. It is, in fact, an expanding forest, with growth being stimulated by the excess CO_2 we pump into the atmosphere. So it seems to be on our side for now. With trees getting bigger and stronger, there is around a net intake of 5000kg of carbon per hectare,[i] meaning the 400 million hectares of forest in total could be taking in as much as two billion (million million) tonnes of carbon every single year.

However, research from the savannah area east of the Amazon basin has discovered that the length of the dry season is the crucial factor determining the development of the rainforest. The savannah to the east of the Basin and the eastern part of the Amazon rainforest receives more or less the same amount of rainfall every year, so why isn't the savannah just more rainforest?

In the savannah the dry season lasts for half the year, but in the rainforest it lasts for merely four months. The longer savannah dry season causes it to catch fire on average twice per decade, whereas the rainforest remains so moist that it never catches fire at all. These fires actually prevent the savannah from turning into rainforest itself, and when you put two and two together you realise the problem. If the rainforest had a dry season as long as the savannah it would likely begin to burn, and if the fires were as frequent as those in the lands to the east, the rainforest would turn permanently into savannah. Ultimately, the Amazon rainforests would stop being our lifeline - the great CO_2 absorbers that protect us from ourselves - and start being huge sources of carbon dioxide instead. Billons of tonnes of CO_2 would burn into the atmosphere over a matter of years, and all our excess CO_2 from the early industrial revolution onward, which has been absorbed by the trees and forest of the Amazon, would be released at once.

The biosphere as a whole is likely to become a source of carbon rather than a sink in the near future. In July and August 2003, Europe's forests and fields hiccupped about 500 million tonnes of carbon back into the atmosphere – roughly double what the continent emitted due to burning fossil fuels over those two months. Soon, the entire land vegetation of the planet will tip from sink to source, and according to Peter Cox at the Centre for Ecology and Hydrology at Winfrith in Dorset, the great tip could arrive as early as 2040.[ii] In fact, this is only Cox's 'best guess' and the point could arrive much sooner than that.

This is not only disastrous for us whenever it happens, but it is disastrous for us now. To help meet Kyoto targets, countries are allowed to include the amount of carbon taken in by their vegetation – some nations are pushing for their targets almost completely by planting new forests. But, since nobody has really accounted for the tipping point in vegetation sink-to-source, the sums could be disastrously wrong. A recent survey of vegetation in Britain – the only one in the world so far – has revealed that this tiny island is releasing about 1 percent of its carbon store back into the atmosphere annually. The total biosphere release is around 13 million tonnes a year[iii] – equal to the amount of CO_2 that the UK saves every year as part of its Kyoto targets!

Climate models have only recently begun to look at feedback effects such as these. Until only a few years ago, the models looked at temperature change within the range of human CO_2 emissions, but we also have to consider how Mother Nature will emit CO_2 herself once we've pushed her too far. And the models are showing that warmer sea temperatures in the southern Pacific will result in less rainfall and a longer dry season for the Amazon. At current rates of emissions, models also show that this flick of the switch in the Amazon will occur around 2050, so we have very little time to act to stop this happening.

A second major area where Mother Nature could unleash a wave of greenhouse gases is in the frozen peat bogs of Siberia. Rising temperatures are beginning to thaw a vast expanse of western Siberian permafrost spanning a million square kilometres – an area the size of France and Germany put together.[iv] The area is the world's largest frozen peat bog and formed to its present size around 11,000 years ago, just prior to the pleasant Holocene. Scientists fear that, if it melts completely, millions of tonnes of methane will be released into the atmosphere – the gas twenty times more potent than CO_2. Ever since scientists first began identifying tipping points in the climate the melting of the Siberian peat bogs has been a scenario they have feared. The feedback loop caused by this event has the potential to be one of the most threatening of all the tipping points so far identified. Unfortunately, the discoverers of this unprecedented thaw, Sergei Kirpotin of Tomsk State University and Judith Marquand of Oxford University, caught the problem at least (they suspect) four or five years after it had begun. During their visits to the region, the researchers found that the usually barren, frozen landscape was now dotted with lakes up to a kilometre across, and vast areas of mud.

Records show an average temperature rise in western Siberia of 3°C over the last 40 years or so – the highest anywhere in the world. Once permafrost is melted only a long-term cooling of the climate can refreeze it. Additionally, when permafrost thaws and turns into ordinary ground, it soaks up more thermal energy from the sun (like the sea does when ice shelves melt) and this accelerates the rate of permafrost thawing.

Methane seeps out of Siberian peat bogs naturally, but the permafrost traps most of what the peat bogs produce, and barely anything escapes at all. Hydrologist Larry Smith, from the University of California in Los Angeles, estimates that the west Siberian peat bogs could hold as much as 70,000 million tonnes of methane, around a quarter of all methane stored in the ground around the world. Though it is highly unlikely it will all melt at once, even a gradual seeping of the gas will double atmospheric methane levels by 2100, and grant us between ten percent and 25 percent increase in further temperature rises.

Worldwide, frozen methane deposits are known as 'methane clathrates' and the Siberian peat bogs are one such example. Clathrates contain far more methane than you would find in any of the gas' surface sources, such as livestock and wetlands. All in all, there could be as much as ten million billion tonnes of carbon lurking in methane hydrate deposits around the world. Though most of this is probably too far below the ocean to ever disturb us, we still have to consider the content at the Arctic. Just how much carbon is lying around the Arctic? It could be in the tens or hundreds of billions of tonnes, and since there is only a mere 5,000 million tonnes of carbon in today's atmospheric methane, the greenhouse effect could be boosted significantly with only a slight melting of Arctic stores.

Though scientists predict that it will take quite some time for temperatures to rise high enough around permafrost and in deep oceans to cause very large amounts of the clathrates to melt, we have to consider that what we do today is what we pay for tomorrow. The delay in the climate will cause the peak concentrations to be reached well after the peak emissions. In other words, even if we stopped carbon emissions growing by 2050, we will not see concentrations stop rising until around 2200 at the earliest (Figures 10.1 and 10.2); hence, though we may not witness clathrates melting in large quantities before 2050, if carbon concentrations don't peak until 2200 then those clathrates are almost certain to melt at some point over the next couple of centuries.

Methane is short-lived in the atmosphere, lasting just a decade or two compared with CO_2, which lasts a century or two. But its potency makes it important. Whilst carbon dioxide pounds away monotonously at the climate, methane molecules prefer to jump in out of the shadows, unleash one almighty punch, and then scuttle off again. More and more methane results in more and more almighty punches - it's only a matter of time before the climate begins to hit back.

Future Temperatures?

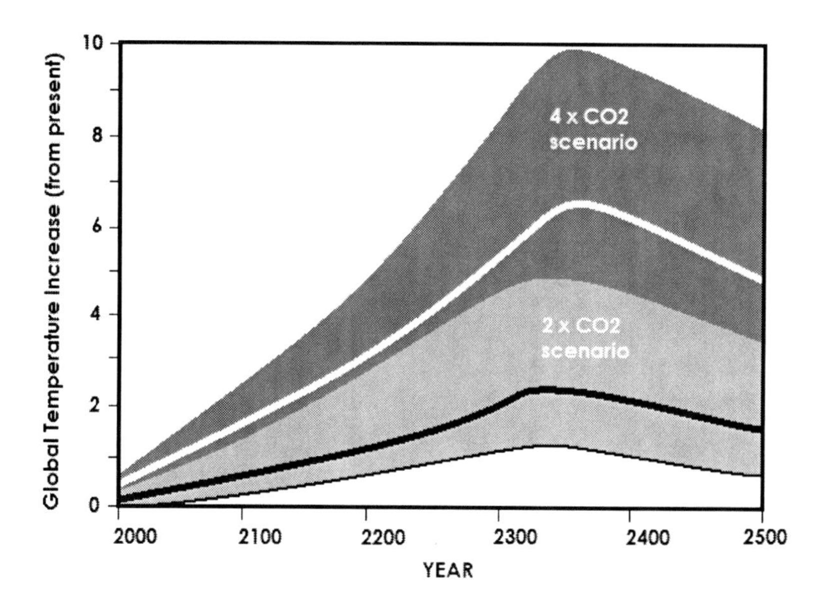

Figure 10.2 – Presuming CO_2 concentrations peak around 2250, the estimated range of the doubling and quadrupling scenarios can be shown.

We can now produce the third and final graph of our series, projected temperature rises over the next 500 years in the $2xCO_2$ and $4xCO_2$ scenarios. Figure 10.3 (above) shows this for us, though there are quite large degrees of error for both lines. For example, the chart shows the $2xCO_2$ peak reach 2.5°C in the year 2350, but this could be as low as 1.5°C in reality, or as high as 5°C. A warming of 1.5°C would be quite fortunate for us, though still significant, while a 5°C warming would produce a very different world than the one we know now.

For the $4xCO_2$, the minimum rise could be around 2.5°C, or even as high as 10°C - scenarios barely comprehendible. This chart presumes that peak *CO_2 concentration* of both scenarios will occur during the twenty-third century, so temperatures will peak during the twenty-fourth century.

Under a $2xCO_2$ scenario, atmospheric CO_2 levels will pass 560ppm – concentrations not seen for around seven million years. A $4xCO_2$ (1120ppm) world has not existed since 50Ma, or probably even around 100Ma during the Cretaceous. By increasing our levels of CO_2, we are effectively sending our planet into its past, retracing a journey in only a few centuries that took tens of millions of years to pass. But before we just look at past climates and assume our future will match one of those, we have to remember that some parts of the climate may respond more quickly to changing temperatures than others. The pulse of future CO_2 levels that we see in Figure 10.1 will last only a few hundred years, whereas ice sheets usually only respond after climate changes lasting several thousand years.

Therefore, when we model future climates, we can only use past climates for reference points, not certainties. In fact, the world of the future will be a strange place, where the atmosphere, vegetation, land and ocean surfaces all respond to climate ups and downs with speed, but the ice sheets drag their feet. This degree of disequilibrium with the climate has not occurred in Earth's history before, so all our present predictions are heavily focused on our models rather than our records.

The tectonic-scale cooling of the planet that we have seen over the last 55 million years should continue on well into the future, but the rate will be so slow (around 1°C every ten million years) that it just wouldn't matter for humans in this millennium. Equally, over the next thousand years, the impact of orbital changes is likely to be too small to bother with, hardly counteracting the effect of a $2xCO_2$ or a $4xCO_2$ world.

So can Gaia save us? Surely life will counteract the changes we humans make, as it has done in the distant past, turning a snowball world into a green and pleasant land? As it turns out, Gaia has not got the power to counteract our emissions simply because we are pumping them out too fast. Nature can only influence carbon dioxide over a very long time period, by expanding or shrinking its forests, adjusting the amount of phytoplankton in the sea, and so on. If we were releasing carbon over many thousands of years then Gaia would be able to keep up and make its necessary adjustments; but we're not. We are releasing carbon at a rate that no natural component of the system can counteract, and we are pushing carbon levels higher than any natural system would want. In fact, the levels could grow so high, and conditions change so much, that life itself cannot adapt or evolve (processes that take millions of years) - forcing much of it into extinction and leaving only the hardiest of organisms. Who knows how many plants will be left at the turn of the next millennium, or how much phytoplankton will be left floating around the oceans. Maybe not enough.

Human-induced climate change is a long-term problem for this planet. Reports in the media, or in wider literature, rarely give climate predictions beyond the year 2100. Not only is this quite pointless from a scientific point of view - cutting off the warming trend before it even peaks - but it also exposes the fact that society is quite self-absorbed; if it affects us and our children then it's a problem - do we really care beyond that?

We have to radically rethink the way we look at climate change. If we don't do enough to stop it then it will not be a problem that will pass in the next hundred years. This is a problem that will be with us, and our descendants, for hundreds, if not thousands, of years to come. That is why climate change is a big deal, because we might soon find ourselves in the footsteps of the polar bear, on a road to extinction.

Too Much Water

The ice sheet over Greenland is beginning to show signs of great stress under the average global temperature rise of 0.6°C already experienced since 1900. As a result, there are pools of meltwater sitting on the surface of the ice sheets that are the same as those that were found over the Larsen-B ice shelf before it suddenly collapsed in 2002. Their appearance over Greenland ice signals the beginning of its demise, an occurrence that could have devastating implications for human civilisation.

On ice shelves, meltwater pools sink down and down until they reach the ocean below, and make the ice look like Swiss cheese, at which point it becomes fragile and breaks up. On land ice, the meltwater is equally problematic. When enough meltwater puddles collect together, and form a large lake, the surface water begins to seep down through the ice in greater amounts. This process can carve deep gashes into the ice sheet itself. Once the water reaches the bedrock below, it is trapped and builds up. Enough build-up of liquid and the ice sheets actually begin to 'float' above the bedrock itself. Eventually the ice will break; forming huge cracks hundreds of miles long in some places, before starting its 'slide' towards the sea. Since the turn of the century, the natural crumbling and melting of the glacier has begun to accelerate to rates never seen before. Figure 10.3 shows Greenland in 1992 and 2005, with the areas of summer melting shaded in red; the comparison is startling.

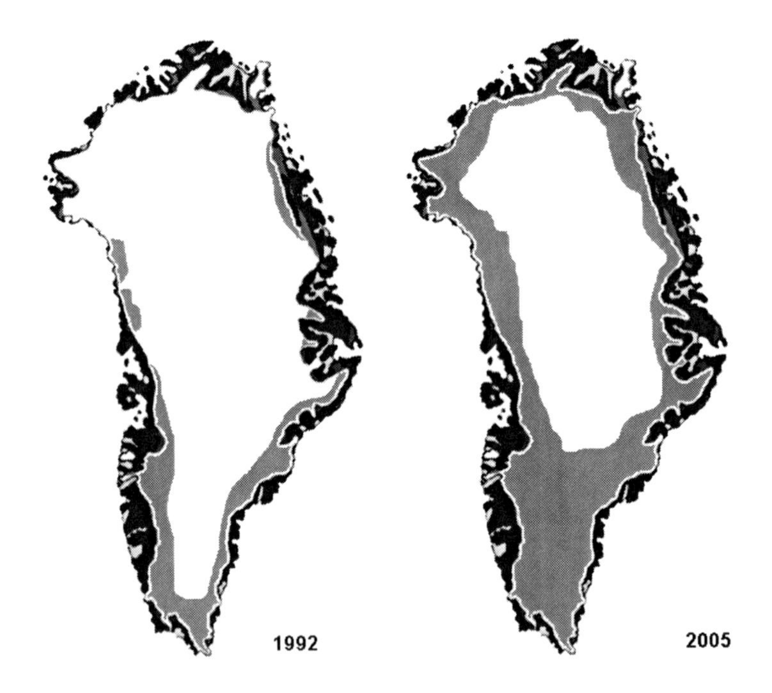

Figure 10.3 (above) – The rate of melting of the Greenland ice sheet is staggering. As the years pass the trend continues, and scientists are constantly readjusting their predictions. How long before the entire ice sheet melts away in the hotter months?

The consequences are even more startling. Though it is still quite unlikely that the Greenland ice sheet will melt completely by 2100 (in winter months), it will certainly do so by the second half of next century if carbon emissions are not halted. A combination of partial melting in the Greenland and the Antarctic could raise the sea level by more than three metres by 2100. According to the US Geological Survey (USGS) complete melting of the Greenland and Antarctic ice sheets would lead to a sea level rise of 80 metres, whereas melting of all other glaciers could lead to only a half metre rise.[v] Melting on this scale is very unlikely, even within the next 1000 years, but it's important to understand the likely *reduction* in these two ice sheets and the consequences. Even a sea level rise of just ten metres would flood around 25 percent of the US population, mostly those living on the Atlantic and Gulf coasts.

Over the last 100 years, the sea level has risen by between 10 and 25cm. Some suggest that the warming and expansion of ocean waters could account for about 2-7cm of the observed sea level rise, while the melting of mountain glaciers could account for 2-5cm.[vi]

And this is where it gets confusing. When scientists make predictions about future sea level rises, their figures are often very different from one another. Indeed, throughout this book we have talked about the West Antarctic ice sheet and the Greenland ice sheet containing 6m worth of water each, 'mainland' Antarctica containing around 60m worth of water, and mountain glaciers holding half a metre worth – but what is going to be the real change?

When I came to research this question I found myself lost in a sea (excuse the pun) of contrasting figures for future sea level. Consider this report from the Food and Agricultural Organisation of the United Nations (1998): "By the worse case scenario, global mean sea-level is expected to rise *95cm* by the year 2100, with large local differences due to tides, wind and atmospheric pressure patterns, changes in ocean circulation, vertical movements of continents, etc." Then there is the figure from the UK Met Office predicting a *41cm* rise by 2080. The differences continue with other reports. Whose figure do you use?

In 2006, a report by NASA alerted the world to the idea that the Greenland ice sheet could be completely melted during summer months by as early as 2040, let alone by 2100. Not only does this suggest a worse 'worse case scenario' than the FAO report of 1998, but it also blows all other predictions out of the water, so to speak. Could we really see a sea level rise of several metres by as early as 2050? The important thing to remember is that sea levels *will* rise and whether the polar land ice melts completely or only around the fringes, it is worth remembering that they *will* melt, as they are already doing today.

Today there are many places on earth that are just begging to be washed away by the sea because they are simply too low, flat, or already flooded as it is. It is an unfortunate coincidence of nature that a species so reliant on coastal metropolises, both today and historically, should be the very species that causes the oceans to rise against them.

We love the sea and always have, though in our world of cars, planes, and nine to five societies it is rare that many of us even consider the sea, it remains one of our chief resources. Consider that many of the largest cities on the planet lie on a river, or near the coastline, or even both: New York is one example. It is estimated that around four out of every ten people on Earth live within 100km of the coast. Nearly seven out of every ten live within 400km of the coast. Every single one of these people will be either directly, or indirectly, affected by sea level rises, meaning that if they are not flooded themselves then their resources will be strained by an influx of people who have been, and the infrastructure of their region is likely to be fragmented and destabilised.

Looking at the Mediterranean we can see that the importance of a direct access to sea and coast is greater for some countries (Italy, Greece, Algeria, Albania and the Gaza Strip) than for others, which may have a stronger inland economy by comparison (France, Spain and Turkey). In Greece, all major industrial and agricultural centres are coast-related, and maybe as much as 90 percent of Greeks live within just 50km of the coast. Though Greece is not a strictly low-lying land a sea-level rise of greater than seven metres will probably destabilise much of the economic foundation of the country and force many of its coastal settlements to be flooded.

We can see the potential for chaos increase greatly when we look at other countries with a high dependency on coastal areas. Though the Nile Delta represents just 2.3 percent of the area of Egypt it provides 46 percent of the total cultivated surface and around half of the population. The northern coastal land of Egypt that lies below three metres above sea level (everything from the sea up to a few miles inland) contains the big cities of Alexandria and Port Said, holding about 20 percent of the population, 80 percent of port facilities, 60 percent of fishing and 40 percent of industry. A sea level rise of just one metre could affect six million people in Egypt with up to 15 percent of agricultural land lost.[vii]

In terms of sheer numbers of people directly afflicted by sea level rises, China (72 million) and Bangladesh (71 million) will be the worst off, according to some sources.[viii] Of all the nations afflicted, the range can be quite small (0.3 percent in Venezuela) to devastating (100 percent in many of the small island nations, such as Kiribati).[ix] Only a one-metre rise will cover around three percent of land surface on the planet, and around 30 percent of cropland will be lost.

CASE 1: At the Mouths of the Ganges

Where the mighty River Ganges meets the River Jamuna, and floods into the Bay of Bengal, is the largest area of low-lying coastal land in the world, and heavily populated. The potential flooding resulting from a 14-metre sea-level rise (loss of Greenland and West Antarctic ice sheet) would devastate Bangladesh, submerging most of it permanently under water, including its capital city Dhaka.

The devastation would not merely be limited to the land covered by the sea – the economic damage and the stress placed on the rest of the country would also be devastating. Today the population of Bangladesh is around 150 million people, and by the end of the century this could be much higher. If around a hundred million people are displaced by the encroaching sea they will exert an incalculable pressure on the part of the country remaining above sea level - plunging the region into massive difficulties in terms of food production and employment. Encroaching seawater inland will also affect fresh water supplies, creating a massive fresh water shortage. It is likely that many of the refugees will flee west into India, which is already one of the most heavily populated countries on the planet. The consequences of up to a hundred million landless people escaping into a country already bulging with people do not bear thinking about.

According to a report compiled by the US Pentagon about future impacts of climate change, increasingly volatile storm surges could cover up to four tenths of Bangladesh by the middle of the century.[x] The subsequent coastal erosion would render much of the country uninhabitable. But the delta area also includes Calcutta, an Indian city with a total urban population of around 13 million people. These people will also need to find a new life somewhere inland. Additionally, the Indian cities of Mumbai and Madurai, with a combined present-day population of just under 20 million, will also find themselves severely flooded, as will Karachi, the most populated city in Pakistan.[xi]

We should also be careful not to forget that Yangon (Rangoon), the capital of Myanmar (Burma) with a population of around five million, will also be flooded, along with most of the surrounding region. In an IPCC special assessment, it is also stated that low-lying areas in Indonesia, Vietnam, Thailand, the Philippines and Malaysia will also lose lots of land.

CASE 2: East Coast United States

With the United States covering a wide geographical area, including extensive coastlines to both the east and west, it will undoubtedly suffer a great deal under any scenario of sea-level rise. The most vulnerable part of the US to flooding is the southern state of Florida. Much of this state is low-lying and a large proportion of the south is swampland. Additionally, there is a series of small islands to the very south of the mainland (the Florida Keys) that are well populated and already suffer from increasingly intense hurricanes almost annually.

Under a seven metre sea level rise (equivalent to losing either the Greenland ice sheet or the West Antarctic ice sheet, of half of both) there would be a large loss of land to the south of the state, including the Keys, with a loss of the unique Everglade swamp system and its ecosystems. There would also be encroachment into cities such as Miami, Tampa, Palm Beach, Mobile (Alabama) and Galveston (Texas). With a fourteen-metre rise, there will be additional flooding in and around all of the aforementioned cities, and also Jacksonville, the John F. Kennedy Space Centre at Cape Canaveral and, additionally, the Bahamas. Just a two metre sea level rise will flood New Orleans and vast areas of the Mississippi delta region. A fourteen metre rise will render the most populated section of the state useless, pushing the coastline back about two hundred kilometres in some parts and flooding urban and suburban New Orleans – an area of around two million people.

Perhaps the world's most well known city, New York, finds itself at the centre of the sea level problem. The five boroughs of New York City, plus a large portion of New Jersey, will be flooded - rendering the entire area unrecognisable and sending what many see as the 'capital of the world' into a watery grave. Roughly 80 percent of urban New York City will be lost – affecting more than ten million people.

CASE 3: Europe

Starting in North West Europe, we automatically see dramatic flooding taking place under a seven-metre rise. In this scenario we stand to lose the majority of the Netherlands (roughly 16 million people out of the population of 16.5 million will be displaced), a large part of Northern Belgium – including Gent, Antwerp and Brussels – and a sizeable chunk of eastern England. In total, this numbers about 18.5 million landless people.

Of course, under a fourteen-metre rise the situation is worse. Large areas of flooding would stretch from South West Yorkshire and North Cambridgeshire, Northern Belgium, and large parts of German and Danish coast. Perhaps the biggest difference between the scenarios is the threat suddenly faced to Greater London with a fourteen-metre rise. No doubt, current flood defences on the River Thames will be vastly improved upon over the coming century as sea levels gradually creep higher and higher. However, by the time sea level increases by fourteen metres (if indeed it does) then it would be uneconomical to keep replacing flood defences with larger and larger versions - and it may not do much good with seawater encroaching across the land as well as just up the river.

London today has a population of seven million people, with up to 12 million in the Greater London area. The economic stress placed on the UK to constantly protect its beating heart, as well as the economic stress in gradually having to abandon its capital city, is only rivalled by the stress placed on the future population.

Perhaps the city at most risk of rising sea levels anywhere in the world is the northern Italian city of Venice. Already a city surrounded by water, and with more canals than roads, Venice is almost certain to fall victim to future climate change before many others. However, loss of just one of the ice sheets, and the subsequent seven-metre sea level rise, will result in a large part of the whole region being flooded. This includes the area around Laguna Veneta, the Po Grande river delta, Ferrara, Ravenna and the northern Adriatic coastline, including Trieste.

CASE 4: Small Island Nations

For small island nations, climate change is the number one issue and there is no number two. In fact, it is such an important issue that these nations already have to deal with the consequences. The combination of intense and frequent tropical storms and sea level rise will, and indeed already does, threaten around forty low-lying island nations around the world, and is beginning to force evacuations and abandonment. In June 2000, New Zealand offered sanctuary to inhabitants of the tiny island of Tuvalu if the sea permanently flooded the coral-atoll. Kiribati has already lost several of its lowest islands to sea level rise.

Other potential victims of sea level rise, even just a couple metres, include amongst the many: Malta, the Lesser Antilles, Réunion, Mauritius, the Seychelles, the Maldives, the Azores, the Canaries, the smaller Philippine isles, the Ryukyu Islands, Tonga, Fiji, Vanuatu, Tokelau, the Midway Islands and the lower Hawaiian Isles.

Sea level encroachment may drive populations further inland in countries like the US, Britain, Italy and India/Bangladesh, but with all of these smaller islands there is simply no inland to retreat to. Most consist of only a single island. Some, such as the Maldives, Bahamas and the Marshall Islands, rarely exceed 2-3 metres above sea level, and others consist of a variety of tiny atolls that only dream of reaching such heights. Even if an island reaches much higher above the sea, encroaching waters may reduce the actual land area to a tiny fraction of its original, practically forcing large quantities of the population to evacuate on sheer land need alone. And then there is the effect of salinization of fresh water supplies on islands that do survive flooding relatively intact.

As a result, governments will have to find new permanent residence in other countries. Where organised evacuations are not carried out to a specified location there could be a widespread outbreak of people taking to boats and aiming to seek refuge upon arrival in the first country they come to.

Destinations of boat people will certainly include the United States and Australia – two countries openly hostile to economic refugees.

CASE 5: Other areas

There are several other sites vulnerable to sea level rise that cannot be ignored, existing all over the planet in both rich and poor regions. Populated centres certain to be at risk from a 14-metre sea level rise include: Alexandria, Port Said, Suez and Cairo in Egypt; Tel Aviv, Haifa, Gaza and Beirut along the eastern Mediterranean coastline; Al Basrah and Al Najaf in Iraq; and Dubai in the United Arab Emirates. Of these places, Cairo is currently home to over eight million people, the Gaza Strip 1.5 million, and Beirut and Dubai both 1.2 million. Taken into account the fact that almost the entire country of Kuwait would be rendered uninhabitable, the number of people affected - not assuming future population growth - could be around 18 million directly.

With Asia being the most densely populated continent on the planet, it is unfortunate that this is likely to be the continent suffering the most from sea level rise. Flooding around the East China Sea, and the potential flooding of the Hebei region of China and the region around the Chang Jiang (Yangtze) river mouth, are all major concerns. Population centres include Shanghai and the capital Beijing, with present day city centre populations of ten million and eight million respectively. Flooding will also occur in Osaka and Tokyo (in Japan) and the southern Chinese city of Hong Kong.

According to the IPCC assessment, a mere one metre rise in sea level will seriously threaten most of Japan's coastline, on which about half of Japan's current industrial production is located. There are also larger areas of China, beyond the major cities, that are at risk of either flooding or salinization of water supplies. In total, again using only present day populations, around 100 million people could be directly affected by ocean waters.[xii]

Other coastal cities around the world will suffer under ice sheet melting – including Sydney, Los Angeles and San Francisco. All of these cities could be saved with flood defences, and populations in these places may have declined by the time sea level rises so high, but the *symbolic* value of these cities being faced with such a threat is extremely important.

Sometimes Change is Bad

The story of future climate change is likely to be one of short, sharp bursts in temperature, arriving one after another like falling dominoes. Such bursts will come with the flicking of several key switches in the climate system, such as the shutting down of the Great Ocean Conveyor, the melting of methane clathrates. It is highly unlikely that the world will gradually shift from one state to another over a long period. Instead it is likely to take several small jumps in one direction that could constitute one giant climate leap, altering the very foundation of the planet.

The first point reached due to future climate change will likely be when the planet's surface warms to a point where it accelerates the hydrological cycle (evaporation, precipitation and runoff), causing temperatures to rise higher still. Water vapour is a potent greenhouse gas, and would further trap heat at the surface. With more evaporation, we are likely to experience drying of forests and grasslands, where biodiversity is high and farmers graze animals and grow grain. Plant life could then either burn or simply just wither and die, but either way they transform from carbon sinks into carbon sources, not only leading to higher surface temperatures but also to fierce and uncontrollable forest fires. Additionally, the melting of glaciers, sea ice, land ice and permafrost will lead to more of the sun's rays being absorbed and less reflected, further compounding the warming.

The consequences of a disruption to the ocean conveyor was examined in the Pentagon report of 2003, in which a 'what if' scenario was run. The report suggested that complete conveyor shutdown would plunge Europe into ice age conditions, making the north west of the continent more like present-day Siberia. In China, it predicts food supplies will be greatly hindered by unreliable monsoon rains, causing chaos and "internal struggles, as a cold and hungry China peers jealously across the Russian and western borders at energy resources."[xiii]

The biggest problem in the Near East will be water supplies. Melting of glaciers will increase, leading to greater flows of fresh water in some river systems, lasting for a few decades until flow is then reduced again with the complete loss of glaciers. It's a case of nature pulling the plug on fresh water in the region – catch as much of it as you can while it rushes away, because one day there'll be none left.

In the northeastern United States the report predicts that growing seasons will be shortened dramatically and become less productive, due to colder and drier weather trends. It points out that change towards a drier climate is particularly prominent in the southern states. Overall, world agriculture is likely to suffer from the length of the growing season being cut by 10-25 percent, and fishing industries that have rights to fish in specific areas will suffer greatly with mass migrations of their prey. According to the IPCC report, agriculture contributes as much as 20-30 percent of Gross Domestic Product in sub-Saharan Africa, and 55 percent of the total value of African exports.

The human cost on population is perhaps one of the least discussed parts of future climate change, perhaps because we don't like to envision our species' dominance on this planet being questioned. But that is exactly what is going to happen with climate change. As we've seen by looking at booming populations and agricultural land-use stresses, we are already riding dangerously close to the carrying capacity[1] of the planet, and that is with the majority of the people in the world having a poor standard of living. If we were to give everyone on the planet the same standard of living as the Rich North we would need nearly three planets' worth of resources. The message from this is that either our demands have to fall drastically or our population does.

With climate change, carrying capacity will be reduced, save some miracle technological solution that everyone seems to be waiting for. After all, human civilisation began with the calming of the Earth's climate. As climate change lowers the world's carrying capacity, it is almost inevitable that more global conflict will occur over food, water and energy supplies. As the Pentagon report says: "it seems undeniable that severe environmental problems are likely to escalate the degree of global conflict."[xiv] A sad sort of justice would then dictate that with war and famine sweeping the planet, populations would decrease, eventually restoring re-balance with the carrying capacity.

The Pentagon report rather gloomily predicts that a shut-down of the ocean conveyor in the near future will push humanity back into its darkest days as a species warring with itself over resources and land. "The United States and Australia are likely to build defensive fortresses around their countries

[1] The estimated number of people the natural world can support considering social and economic systems and technological know-how.

because they have the resources and reserves to achieve self-sufficiency... Borders will be strengthened around the country to hold back unwanted starving immigrants from the Caribbean islands, Mexico and South America."[xv]

Though the Pentagon report offers us only an insight into a particularly extreme and abrupt climate change scenario, there is no reason to simply brush it aside. Warfare has often been carried out throughout history over natural resources, as nations outstrip their carrying capacity; there is no reason to believe that a future of resource scarcity, failing water and food supplies will not lead to internal and external disputes, and mass migrations of refugees.

Today there are over ten million 'environmental refugees' forced away from home by drought, erosion, urban sprawl, loss of farmlands, and pollution. It is the single biggest class of homeless and landless people - save only for Palestinian refugees living in camps in places like Jordan. This figure will rise dramatically if greenhouse warming kicks in violently. We have to realise that global warming will not give us a future of wealth and variety. It will cause sea encroachment around low-lying coastlines, landslides in the mountains, soil erosion and flooding in valleys, drought in some agricultural land, forest fires, increased sea acidity, more tropical storms such as hurricanes and typhoons, and certainly a lot of refugees. In effect, this leaves us with virtually no place to go that isn't negatively affected by climate change, because our mountains are no safer than our valleys, and temperature rises are expected to be so rapid that biodiversity is seriously threatened even in the most remote locations.

---

[i] Hydrogen NOW! Journal H2 – Article 1.
http://www.hydrogen.co.uk/h2_now/journal/1_global_warming.htm, Accessed 5th Sept, 2005

[ii] Pearce, F. (2006). *The Last Generation*. Transworld Publishers, London. Pg 106.

[iii] Ibid. Pg 107.

[iv] Sample, I. 'Warming hits 'tipping point', *the Guardian*, October 11th 2005.

[v] USGS webpage, http://pubs.usgs.gov/fs/fs2-00/, Accessed 9th Dec 2006.

[vi] GRIDA webpage, http://www.grida.no/climate/vital/19htm, Accessed 9th Dec 2006.

[vii] Zwick, A. (1997). 'Monitoring and assessment of recent research results on global climate change with a special section on sea level rise and joint implementation' in: *Climate change research and policy*, the Commission of the European Communities, Update N. 9. EU Joint Research Centre, Sevilla, EUR17303EN. Pg 76

[viii] McMichael, A. J., Haines, A., Slooff, R. & Kovats, S. [eds], (1996). *Climate Change and Human Health*. WHO, Geneva. Pg 296

[ix] Gommes, R., du Guerny, J., Nachtergaele, F. & Brinkman, R. (1998). 'Potential Impacts of Sea-Level Rise on Populations and Agriculture', Sustainable Development Department (SD Dimensions – FAO) webpage, http://www.fao.org/sd/EIdirect/EIre0045.htm, Accessed 3rd Jan 2007

[x] Bunyard, P. (1999). 'A hungrier world,' *The Ecologist*, March/April.

[xi] Schwartz, P. & Randall, D. (2003). 'An Abrupt Climate Change Scenario and Its Implications for United States National Security,' http://www.grist.org/pdf/AbruptClimateChange2003.pdf, Accessed 12th Oct 2006.

[xii] Watson, R. T., Zinyowera, M. C., Moss, R. H. & Dokken, D. J. (1997). 'The Regional Impacts of Climate Change: an Assessment of Vulnerability,' Special Report of the IPCC Working Group II, published of the Intergovernmental Panel on Climate Change. November 1997.

[xiii] Schwartz, P. & Randall, D. (2003). 'An Abrupt Climate Change Scenario and Its Implications for United States National Security,' http://www.grist.org/pdf/AbruptClimateChange2003.pdf, Accessed 12th Oct 2006. Pg 13

[xiv] Ibid. Pg 14.

[xv] Ibid. Pg 18.

PART 4 - the Quick Guide on How to Save the World

Chapter 11 – Technological Solutions to the problem

'Being defeated is often a temporary condition. Giving up is what makes it permanent.' – Marilyn Von Savant, American columnist.

While we still burn fossil fuels, we are causing climate change - storing up the damage for the future, when the delay in the climate system unleashes it upon us. The vast majority (61 percent) of our greenhouse gas emissions come from energy production. Of the energy demanded, around 40 percent is for electricity and heat production, 22 percent for transport and around 38 percent for industry. Therefore, the single best place to start when we want to tackle greenhouse gas emissions is with our energy production and in particular our electricity. So how do we do this?

Infinite Energy

Energy is all around us in the natural world, and virtually all of it doesn't involve burning anything. We see it in the movement of the seas, the flow of rivers and waterfalls; we feel it in the rays of the sun and in the wind; it even exists below ground in boiling hot gases and even in earthquakes. All we have to do is figure out how to harness nature's power for its transformation into our own energy demands.

Perhaps the easiest form to transfer is wind energy. This is because we have had the technology for harnessing wind power for centuries now, going back to windmills that served to grind grain and other such agricultural products. You may even say that we've been harnessing wind power since the earliest civilisations many millennia ago, when humans first took to the water in ships and boats with sails. Wind power is one of the cheapest options for producing electricity and it consumes little energy in construction and deconstruction. There are disadvantages of course: a wind turbine only produces between 0.25 and 2MW of electrical power, and must be located in areas that receive a lot of wind. Because they produce only a little energy per turbine, a wind 'farm' must contain multiple turbines to have any noticeable effect. For this reason, it is better to have wind farms in very remote locations, away from human populations and environmentally sensitive areas.

Wind turbines produce no waste or pollution, require no fuel, are relatively quick to construct, have low maintenance costs, and use little energy from being built to being taken down. Additionally, it is worth remembering that the power output depends on the wind speed cubed; in other words, if the wind speed doubles then the power output increases by a factor of eight. Wind power is slowly being adopted by nations as an attempt to tackle greenhouse gas emissions, and there are already many large-scale schemes around the world.

We can also harness the energy in our seas, in two different ways. The first is tidal power, which makes use of the fact that the level of water in oceans and estuaries rises and falls in predictable patterns. In this way, the rise in seawater can be trapped at high tide behind a dam or barrier and allowed to run out through turbines at low tide, thus creating electricity. Tidal patterns are very reliable and the tides are always being monitored around the world because of shipping and navigational demands.

Best of all, tidal energy harnesses the power of the planet's rotational movement through space, and hence it will never stop happening. When a point on the Earth is aligned with the moon, the gravitational effect lifts the water slightly, producing high tide. The rotation of the Earth moves that point out of alignment, which produces low tide. Far out at sea the tidal increase of the water is less than a metre but in other locations, such as funnel shaped estuaries, large tides exist. This means that the best locations for potential tidal power stations are very close to home indeed, and, though they may be moderately expensive to construct, at least they don't require any sort of construction far out into international waters. It should also give added incentive to nations with long coastlines, especially the UK, USA, China, Japan and Australia. In fact, the UK alone has about half of Europe's wave, tidal and marine current potential.

Harnessing the waves is a very similar process to the tides, though there are many methods for doing so. One of these methods involves channelling wave water up a converging ramp into a raised reservoir on the land. This is necessary so that the water can be slowly released back into the sea through another pipe, which passes through a turbine. It's as if the water is being captured, mugged for all its energy and then released back into the sea bewildered and dishevelled. The second method involves what is essentially a concrete bunker, enclosing a column of air on top of a column of water. The bunker then fills up with water at each wave, pushing and pulling the air through a Wells turbine, which has the ability to rotate in the same direction regardless of direction of flow. There are also other methods involving devices floating out at sea.

Wave power is, after all, just a concentrated form of solar energy. We can use the sun's radiation more directly with two principle methods. One of them is so simple that today it is quite an aged method, whilst the other one still seems quite futuristic.

When early travellers filled black pans with water and left them in the sun until the water was hot, they were using solar power. Today we can attach roof-mounted solar panels to our buildings to provide us with heat. Water circulating through them heats up, and can then be pumped to a main boiler in the building or carried elsewhere. Efficient systems can meet the hot water demands for a whole family – around 70 litres per person per day. In extremely sunny places, like the sunbelt states of North America, such installations can pay for themselves within four or five years by savings on bills. If most systems last for more than 20 years, this gives you at least 15 years of absolutely free hot water. Even in places that aren't so sunny, solar heating can create huge savings.

The more futuristic form of harnessing the power of the sun is something known as photovoltaics. Interestingly, the technique was actually discovered in 1839 by Edmund Becquerel, who noticed that light directed onto the electrodes of a battery cell increased its voltage. In 1954, a team at Bell Laboratories managed to create a solar cell that could convert sunlight into electricity at six percent efficiency. It wasn't long before they were implemented by the space program for both craft and unmanned satellites; since then the conversion efficiency of solar cells has increased to around 25 percent.

A single solar cell produces hardly any electrical output at all, but they can be manufactured so that they can be connected in large blocks and arrays, producing large amounts of energy. Like wind and sea power, electricity can be produced with solar cells with zero waste and pollution during operation, and require no transportation of fuels like fossil fuel power stations do. One of the world's largest arrays of solar panels is Solar One in California, providing

10MW of electricity. Dish systems use a giant dish up to 15 metres in diameter to focus the sun's heat on a receiver. Such systems can produce around 15kW of electricity each. A massive system in Southern California produces about 90 percent of the area's electricity for hundred of thousands of people. Solar thermal plants can be expensive but quick to install.

The potential is huge throughout the world, with around two million settlements without electricity today that are within just 20 degrees latitude of the equator – the zone that receives the most solar radiation on the planet. In the temperate zones at higher latitude, each square metre of a south-facing roof receives about 1000kWh of radiation every year – more energy than we need to heat rooms and water. Though we can only harness part of this total energy, it can still help reduce our dependence on fossil fuels and provide most of the average family's hot water requirements from May to September.

The deserts of Southern California are perfectly suited to solar energy, being both rich in sunlight and... well, deserted. Two Californian utility companies are planning to build a pair of solar power plants that will be far superior to existing facilities, and seriously rival the fossil fuel equivalents. Stirling Energy Systems plans to build one solar plant with the capacity to generate 300 megawatts of electricity, and another to generate 500 megawatts. These plants will utilize a method very different to photovoltaic cells, and instead focus the sun's light onto a Stirling engine, kind of like how household satellite dishes focus the signal onto the receiver. The Stirling engine is also known as the 'external combustion engine' and dates back to 1816 and Scottish clergyman the Rev Robert Sterling. The engine is a safer alternative to the internal combustion engine (whose boilers were prone to exploding) but did not provide the instant power required for transport, like its ugly brother. When built, the new plants will raise the present solar output of the US from about 400 megawatts to about 1200 megawatts. Construction should cover about 18 square kilometres of unused desert with 20,000 dishes. With enough will power, other schemes of similar scale could be placed in many spots over the planet, including the Sahara Desert, Arabia, Spain, southern Africa, Australia and central South America.

A final option also exists for today's renewable energy production, and that is a sector known as Biomass. This is a term used to envelop many sources of fuel which are all around us and can be turned easily into energy. Essentially all sources involve combustion but the subsequent pollution can be much lower than combustion of fossil fuels. Other types involve 'gasification', where the organic matter is converted to a gas by heating it in the absence of oxygen and thus converting to hydrogen, CO_2 or methane. There is also something known as Anaerobic Digestion, whereby organic wastes such as farm slurry or food waste is converted to methane, which can be burnt.

There is the concept of growing crops specifically for the purpose of burning them for energy production (such crops include willow trees which grow relatively quick). Emissions from wood-fired power plants are far lower than coal-fired ones, and transportation costs are much lower since trees can more or less be grown on the doorstep of the plant itself.

However, combustion of anything is still a poor source of energy from an environmental perspective, especially if landscapes are transformed so that 'farms' of trees can be planted. There is also the negative effect of monocultures developing and ruining biodiversity. Ultimately, combustion of biomass may be the lazy way of switching off fossil fuels but it cannot be the final answer.

When a country switches some of its energy production to renewable resources, not only is it reducing its CO_2 emissions, but also it is strengthening its energy security. With a large worldwide energy crisis looming within the next 20 years, any nation with a strong proportion of energy coming from renewables will have the advantage over everyone else.

So what answers do renewable sources have for us in the long-term future? Well, what better way to produce the energy we need than building our very own Sun? The method is called nuclear 'fusion' and employs the same basic elements as conventional nuclear power plants, only instead of breaking atoms apart to release energy, the fusion method forces atoms together - rather like today's nuclear production but in reverse. Nuclear fission (the breaking apart) releases energy from highly radioactive sources by breaking the bonds between atomic nuclei - fusion involves squashing atomic nuclei together, only it uses two hydrogen atoms rather than uranium. This process releases a helium nucleus and a neutron, plus lots and lots of energy.

Fusion has several advantages over the old-fashioned fission process. First, it only requires heavy hydrogen (or deuterium)[1], which can be easily extracted from water, and a bit of super-heavy hydrogen (tritium), which can be made from the reasonably abundant element lithium. Therefore, the fusion reaction requires fuels that are quite cheap and abundant compared with the limited, expensive and highly dangerous uranium used in fission.

Secondly, fusion releases energy on a scale that dwarfs anything we can really comprehend. According to one source, you could supply all the energy needs of the average European for 30 years with nothing more than the lithium in one laptop battery and the heavy hydrogen in half a bath of water. And thirdly, fusion is incredibly safe. Unlike fission, there is no chance of a runaway meltdown as happened at Chernobyl; if you stop applying the fuel, or switch off the equipment, the reaction just stops. Furthermore, it takes barely any fuel at all to get fusion to work – less than a gram of fuel compared with the 250 tonnes used at Chernobyl. And this tiny amount of waste is also much less radioactive than its fission counterpart, applying a nice icing on the cake.

However, despite understanding the physics fully, scientists are finding it difficult to harness the method. Surprisingly, research into nuclear fusion dates back to the 1950s, when scientists thought it would provide reliable power by the end of the century. The problem they have is getting the two nuclei close enough to fuse and controlling the reaction. To do this they need to insert a large quantity of energy into the reaction at first to convert around a gram of the fuel into 'plasma' – the fourth state of matter, (where a gas has been heated superhot, like lightning for example). For the reaction to work there needs to be a doughnut-shaped chamber surrounded by an electromagnetic jacket. This design goes back to 1960s Russia and stops sub-atomic particles within the plasma.

Still, the experts press on with the idea, knowing that in there somewhere lies the key to massive amounts of energy using nothing but water and tiny amounts of lithium. Despite its huge research costs, and the fact that it is still many decades away from being put into commercial use, fusion energy production seems almost perfect, and it is with this perfection in mind that a new generation of experimentation and application is under way.

---

[1] Hydrogen is made up one a proton with an electron spinning around it. Heavy hydrogen has a nucleus of a proton and a neutron.

But this seemingly limitless supply of cheap and carbon-free energy is not so science fiction. In fact, there are 28 fusion reactors around the world smashing matter together at incredibly high-speed on a daily basis. By far the biggest is Jet in Oxfordshire, England, built in 1983 and the first in the world to use deuterium and tritium in 1991. It proved that the process could create energy, but that a much larger scale was needed in order to get out more energy than you initially put in.

In 2006, a seven billion pound project called Iter (International Thermonuclear Experimental Reactor), aiming to build a prototype fusion reactor, was agreed upon by the EU, Japan, China, India, South Korea and the US. This project hopes to provide the scale needed to produce an energy profit rather than a shortfall like at Jet, boasting that it will produce ten times the energy needed to start the reaction. "We will demonstrate scientifically and technically that nuclear fusion energy is a viable energy source," said Akko Mass, one of the scientists working on Iter.[i]

Alas, even the miracle of nuclear fusion has its downfall. Scientists don't think fusion will be ready to provide widespread energy needs around the world any time soon, and some even think it could be 2100 by the time we see it in commercial use. But it is certainly a method worth pursuing for mass application in the future, and its experimental costs are justified reaching into the billions in a world that spends multiple times that amount on space programmes and military budgets. The only problem is that it doesn't really help us in the here and now – what do we do for energy until fusion comes around?

To eradicate the use of fossil fuels completely in future, a brand new system will have to put in its place – a system capable of replacing *current* energy demands and also providing additional future energy demands, which could be double that of today by the end of the century. So what else, apart from fusion power, could we utilize to supply this energy? Some experts are pointing towards an economy centred on Hydrogen. This, you will soon realise, is more of an umbrella idea, where a hydrogen economy will cover everything from energy production to transport, and therefore fusion power is included in that umbrella.

Hydrogen is the most abundant element on earth, and the universe, but it only exists on our planet in various different compounds with other elements, i.e. with oxygen as water. Free hydrogen molecules are too light to hang around in the atmosphere, and instead drift away into space. Therefore, to extract hydrogen from these compounds it takes an initial burst of energy to be put in. Though hydrogen is flammable when it is burnt it only combines with the oxygen in the air and produces water. In the production of hydrogen, you can either reform fossil fuels with a reformer or pass a current of electricity through slightly acidic or salty water (meaning electrolysis can take place in a system at home, producing hydrogen to power a car fuel cell or the home electrical supply).

The first method is, of course, slightly perverse, in that they require the use of fossil fuels as the source of hydrogen for the hydrogen economy. Oil and gas contain hydrocarbons, which can be split into hydrogen and carbon molecules relatively easily using a reformer. The leftover carbon is then expelled into the atmosphere. This may reduce pollution, but it still requires both fossil fuels and CO_2 emissions. Fuel-cell powered cars are almost completely powered by hydrogen taken from petrol using a reformer, simply because petrol is such a readily available fuel. Still, using fossil fuels to extract hydrogen may be a good temporary step to take during the transition period to a hydrogen economy, and is certainly better than doing nothing.

The second method of electrolysis is not a primary source of energy output, and instead requires electrical energy to be put into water to convert to chemical energy. So where will this electricity come from for this 'pure hydrogen economy'? The answer is a tricky one. It would mean that electrical energy would be needed to not only replace existing fossil fuel power stations, but also to pick up the energy tab of transport (cars, planes etc. that all run on fossil fuels today). Existing renewable sources like wind and solar can only take on a chunk of this burden, even with significant investment. So how do we produce this electricity, remembering to avoid an expansion of nuclear fission? The problem remains.

Since hydrogen is pollution-free, renewable, and promises to supply energy for everything from our home electricity down to our automobiles; switching over to a hydrogen economy would be advantageous in many ways. Fossil fuel economies rely heavily on each other, meaning that there is competition to import things like oil and coal, and leaving countries vulnerable to market fluctuations. Hydrogen allows countries to be completely self-sufficient in their energy supply. An eventual switch over from one economy to the other will also produce thousands and thousands of new jobs.

However, being self-sufficient in energy supply is only a good thing in a utopian world. Currently, governments (particularly the US) spend literally billions of dollars every year subsidizing oil exploration, purchasing oil from foreign governments and privately owned corporations, and militarily defending access to oil in the Near East. Is there anybody out there who would actually lose out if countries stopped doing all this? Of course there are, and these people are extremely wealthy and extremely powerful, and not likely to accept a hydrogen economy lying down.

How to utilize infinite energy:

In places like the UK, wind turbines are best placed out at sea where there is plenty of wind and no people, though it's no good plonking them in the middle of very sensitive marine ecosystems. Onshore wind power is also necessary, especially now when the climate crisis is so grim. If we had started to get moving on preventing climate change earlier than we did, then we'd have the luxury of rejecting onshore 'unsightly' wind farms; but we've left it too late, and if we want to cut out fossil fuels as quickly as possible then onshore wind schemes must be implemented as soon as possible. And let's not forget, beauty is in the eye of the beholder: some people actually like the look of wind turbines on a hillside. Consider also the other scars on our countryside: electricity pylons, industrial complexes, open-top mines, new homes etc. These may be disliked but they are accepted because they are needed - can another exception not be made?

In terms of the hydrogen economy, and the problem of where we could get the electricity to produce the hydrogen, the answer could lie in sharing. According to some scientists, by distracting ourselves with fossil fuels and technological alternatives far off in the future, we have been missing out on a vast amount of cheap renewable energy in the present.

Every year, the amount of solar energy reaching Earth is enough to supply world energy demand many times over. The problem has always been in harnessing this energy. The Sahara Desert alone receives the equivalent energy of around one and a half million barrels of oil every single year... for every square kilometre! In other words, potentially there is enough solar radiation in the Sahara to power the whole planet.

However, unlike solar heating or photovoltaics, the best way of harnessing this energy is a method known as Concentrated Solar Power (CSP), which involves focusing the sun's energy onto a pipe of liquid using nothing but mirrors. The liquid can be heated quickly to a temperature of around 400°C, transformed into gas and used to power an old-fashioned steam turbine. This technology is nothing new, in fact we only have to go to the deserts of southern California and Mojave Desert to see that it has been used to produce electricity for years; other CSP plants are even being built in Australia and Spain. The steam can also be stored for long periods for cloudy days and during the night.

It is the fact that this technology is very simple, and already in use, that means there will be virtually no research needed to set one up in the Sahara, or the deserts of the Near East. And the two scientists pushing for the Sahara/Near East scheme, Dr Gerhard Knies and Dr Franz Trieb, calculated that we would only have to cover half a percent of the world's hot deserts to provide enough electricity for the entire world.

So what's really revolutionary about this idea? Two reports, prepared for the German government, focus on the value of sharing energy between North Africa, the Near East and Europe, using a kind of super-grid of low-loss, high voltage direct current (HVDC). This would require collaboration between countries of North Africa, the Near East and Europe. Virtually all countries have in place a national grid system for dispersing electricity from power stations to the consumer; all that would be required is to join these national grids up to form a single 'super-grid' and to convert them to DC cables (direct current) rather than their current AC cables (alternative current). Using DC reduces the energy loss over long distances, meaning between North Africa and The British Isles the energy loss would only be around ten percent. Additionally, European countries could put in their own electricity production from alternative sources: Norway's hydroelectric power, British wind power, central European biomass etc. The super-grid could unite the Euro-Arab world in sharing energy, provide all of their energy needs by 2050 with barely any fossil fuels or nuclear energy, and allow a 70 percent or higher reduction in carbon emissions over the period.

Best of all is the fact that hot deserts of the world, such as the Sahara, the Arabian peninsular and Iran, are not suitable for other uses and hence very cheap, but they receive roughly three times the sunlight of northern Europe. The costs of CSP would be equivalent to around $50 per barrel of oil to build a plant, but this cost is likely to fall to around $20 as production of the mirrors reaches an industrial scale. This is about half the cost of using photovoltaic cells and much lower than oil itself, which currently hovers around $60 a barrel.

For additional bonus, the mirrors are quite large and create lots of shaded areas, allowing them to be used for horticulture, irrigated by fresh water created by the power stations. This would add large areas of agricultural land to the world stock, possibly helping alleviate food shortages, especially in North Africa. On top of this, the cold water generated by the plants can then be used in air conditioning systems throughout nearby hot cities, further reducing power demands. It is this triple edge of CSP technology that can boost the overall efficiency to around 80-90 percent.

Taking all three benefits into account, CSP becomes more competitive than even natural gas for producing electricity, and that is before you start thinking about things like carbon emissions. And just because CSP undercuts fossil fuels in the market place doesn't mean that the world's power elite have to miss out on their fortunes. Most oil producing nations have very large areas of hot desert land that they could easily take advantage of.

However, when it comes to energy and resources, countries can often act like spoiled little children in a playground, and the idea of sharing may not go down very well. Governments also seem quite ignorant of the potential of CSP technology, and many advocates of the scheme are worried that politicians could stand in the way of an idea that could save the world.

In the meantime, we should be demanding decentralization of energy production, away from big power plants and towards individual homes and office blocks. We may not be able to provide all of our energy needs from the roof of our own homes using tiny wind turbines or solar panels, but it will certainly help lower the generation needed from power stations. The UK Energy Saving Trust estimates that by 2050 around 30-40 percent of the country's energy needs could be met by decentralized energy.

Green Vehicles

Today, transportation contributes about 13.5 percent of world greenhouse gas emissions. In the UK alone, CO_2 emissions from transportation have risen by 50 percent since 1990. This is hardly surprising when you consider that the number of motor vehicles worldwide grew from 53 million in 1950 to more than 555 million in 2001 – growing by about 100 million per decade since the 1960s.[ii]

Around one tenth of all carbon emissions come from road transport, and we can lay the blame for a lot of that over the last decade with the increasing production and sales of 4x4s. Four-wheel drives like Range Rovers and Grand Cherokees are highly energy inefficient, guzzling more fuel than other car types, and contributing a lot of pointless CO_2 to the atmosphere. Driving a 4x4 in urban or suburban areas is akin to cracking a walnut with a sledgehammer. Additionally, many 4x4 models only do 12 miles to the gallon, which is less than the Model T Ford did 80 years ago.

Electric vehicles have existed in the form of milk delivery vehicles for many years, and it's been nearly 120 years since the first electrical car was built[2]. But milk vans are easy vehicles to make because they require neither great speed nor acceleration, and this is where the electric car falls flat on its face. In today's world, an electric car would only be viable if it could match all three of the following criteria:

* High torque, high-speed motors of high efficiency;

* Motor control circuitry of high reliability;

* High power and high-density batteries.

The first two requirements have been mastered, but adequate batteries pose a real problem to us and prevent electric vehicles from realistically rivalling the internal combustion engine of an ordinary car. Some exotic batteries, made from hazardous or expensive materials, have proven to show improved performance, but only the traditional lead battery has any real commercial potential.

[2] The first successful electric car is traditionally credited to William Morrison of Iowa in 1891, although Scot Robert Anderson invented a crude electric powered carriage sometime between 1832-1839. Electric automobiles represented almost a third of all cars on the roads of Boston, Chicago and New York City around 1900. With Henry Ford's mass-produced Model T – powered by petrol – the rise of the electric car was halted.

Batteries are necessary for electric cars because they must store energy without burning fuel. When we use cars we do a lot of stopping, starting, slowing down and accelerating. This means that we require large bursts of energy to be released at some times and prolonged energy output for times when we need to make longer trips. To be useful, a battery must contain large quantities of energy and be able to deliver it at a rate that may at times be rather large too. At the moment, batteries are poor at doing both for very long periods, and manufacturers are having trouble developing one that can go the distance. In short, the electric car just hasn't got the stamina.

A further difficulty that has prevented this type of transport from taking off is the problem of charging the battery. Throughout the world, there are solid infrastructures in place to distribute petroleum and refuel cars quickly and cheaply. How would you like to spend eight hours to completely refuel your car, rather than just five minutes? In most industrialized countries, people simply do not have the time to let their cars sit idle whilst recharging. Furthermore, it would require a complete overhaul of the supply infrastructure, with recharging stations having to be set up in as many places as petrol stations are at the moment, and there is no evidence to suggest that current electricity supply methods could cope with the increased demand.

Electricity in cars remains best used to supplement the internal combustion engine, in what are known as 'hybrid' cars. Such vehicles usually burn petrol for accelerating only and for cruising along the battery will kick in – essentially reducing fossil fuel consumption rather moderately. Because the batteries charge whilst the vehicle is in motion you don't have to worry about plugging it in and charging it. The market for hybrid cars has gone from being a very small niche to an increasingly important segment. Many major manufacturers are producing hybrid vehicles, marrying the seemingly opposing forces of the combustion engine and a low emission vehicle in sleek, quiet and efficient models. What's more, prices of hybrid vehicles are gradually falling; where once they may have set you back more than £30,000, they can now be bought for considerably lower.

There is also the fuel cell vehicle. If you want to get technical about it, the fuel cell is an 'electrochemical energy conversion device' - converting hydrogen and oxygen gas into water, and in the process making energy. Fuel cells can be applied anywhere that energy is needed, including homes and buildings, in industry and present power stations, but they are particularly suited to cars. However, like the solar cell, fuels cells involve a lot of complicated maths and physics, so for the sake of avoiding both we'll just stick to how it can be applied rather than how it works.[3]

The output of a fuel cell is powerful and relatively efficient, and the only waste product is harmless water. However, there are problems with the fuel cell when it comes to applying it to everyday life. For a start, the fuel cell requires hydrogen and oxygen to work, and though oxygen can be taken in from the air, hydrogen is much less abundant, difficult to store and there just isn't the infrastructure in place to pump hydrogen to homes or filling stations for cars. To get around this problem, scientists have come up with something known as a 'reformer', which turns other gases into hydrogen. Alcohol has been used but the best appear to be methanol, natural gas and propane.

From an efficiency point of view, a fuel cell powered with pure hydrogen can be up to 80 percent efficient at creating electrical energy. With a reformer, this drops down to 40 percent or thereabouts; when you consider that a motor can transform electricity into mechanical power with an efficiency of 80 percent, the whole efficiency of a fuel cell becomes between 24 and 32 percent.

'Poor show' I hear you say, but in comparison, the overall efficiency of a petrol-powered engine is only about 20 percent, since most energy is lost in heat, noise and all the pumps and fans that keep it going. Electric vehicles prove to be around 72 percent efficient altogether, but this can be affected by the origin of the electrical power that must be put into the battery in the first place. For example, if the electricity came from a fossil fuel power station then only about 40 percent of the fuel was converted into power and charging a car has an efficiency of 90 percent. This brings the overall energy efficiency of an electric vehicle down to 26 percent. However, if it gets its electrical power from a renewable source then the absence of fuel at the power station brings the whole efficiency of the car back up to around 65 percent.

The fuel cell is a wonderful technology, which, had it not existed in our present day fossil-fuel-obsessed world, would be an excellent answer to our energy demands. One of the reasons it is held back is because we have a current infrastructure in place for fossil fuels only, and alternative fuels are still marginalised. Still, the fuel cell is beginning to make waves, with mini

[3] For a good explanation of how a fuel cell works visit http://www.howstuffworks.com/fuel-cell

fuel cells being researched for use in laptops, mobile phones, bicycles and other applications. It is also being used today in buses around many cities, and some experimental homes.

A final option for green vehicles is a collection of fuels known as biofuels. The total mass of living matter (including moisture) on Earth could be around two *billion* tonnes - 90 percent of that existing in plants alone. Organic matter may represent only a minuscule amount of the Earth's total mass but it is vital to our survival both as a moderator of climate and as a source of energy. More importantly, it is an energy store that is constantly replenished.

Technically, biofuels include every solid, liquid or gas produced from organic materials that can be burnt, including plants and agricultural, industrial, commercial and domestic wastes. Until around 1800, the history of all fuels was more or less the history of biofuels. The replacement of wood by coal just prior to the Industrial Revolution had been because of a shortage of wood (since most trees had been cut down by then). Considering that around two-thirds of the people on the planet still depend on burning wood for their primary energy supply, this becomes extremely relevant indeed. What is the future of biofuels in a Poor South where populations and energy demands are booming? Will it increase or decrease?

Biofuels can generate heating and electricity, but perhaps their best application can be in vehicles. Biofuels are essentially liquid fuels from energy crops such as rapeseed oil, sugarbeet or vegetable oils. Simple conversion of rapeseed oil can fuel diesel vehicles with little or no modification to the engine or tank. Sugarbeet can be fermented into alcohol and used as a petrol substitute or additive, meaning petrol will be only a minor player in powering your vehicle.

There are drawbacks to this type of fuel, one of the most problematic being that few companies see the market as one to invest in. Even when companies do decide to invest in biofuel production they have to convince farmers to grow the energy crops they need, and a lot of convincing takes a lot of money. Many farmers refuse to get involved because it usually means they would receive no income while the system is being set up and this can take many years – even then, many do not see a secure income over the long-term.

Another problem stems from the fact that governments still insist on taxing this form of fuel – even if they have the rate much lower than oil-based fuels and various incentives for producers and consumers alike – and this means that the margins of profitability are still not really worth major investment by industry. It's hard not to sense a little pressure being applied by those who might lose out in the energy market.

Still, biofuels for vehicles could expand over the coming decade as a sector and, with adequate incentives in major countries like the US, the outlook is good. However, it is doubtful whether biofuels can provide more than a modest amount of our transport energy needs and, though it stands out as a strong transition fuel from the Fossil Fuel Era to the Hydrogen Era, it will probably remain on the fringes.

In the end, the future of alternative vehicles depends not on the efficiency of the engine or power supply, but rather on its commercial potential. People want vehicles that are easy to refuel, travel quickly and last a long time before refuelling again. The next generation of vehicles will be a compromise between efficiency, emissions and practicality. For the immediate short-term, the biofuels may prove the cheapest and most convenient alternative.

As we've seen, any practical alternative to fossil fuels in homes and cars must rival a system of delivery already in place, and therefore it is difficult to imagine any government or industry would wish to ignore these cheap alternatives and go for a complete revolution in energy systems. Instead, we are very likely to see a blurred transition from fossil fuels to non-fossil fuels with an intermediate period of hybrids and mixtures. Will the slow phasing out of fossil fuels actually do very much to curb climate change?

Carbon capture: buying ourselves time

For all this technological advancement to take us away from fossil fuels and into a bright and clean future we have to wait - in some cases this waiting could be for many decades yet. Additionally, the transition period between present day greenhouse emissions to a future of zero greenhouse emissions will obviously see overall emissions decline gradually, but this will subsequently see atmospheric greenhouse *levels* continue to grow. So while we wait for technology to arrive, we cannot just sit around twiddling our thumbs.

Also, aside from the climate change problem, levels of oxygen in the atmosphere appear to have been dropping over the last century or so. According to some research, the earliest years of life on Earth saw oxygen levels in the lowest layer of the atmosphere that could have been as high as 38 percent. The human body may have evolved at a time when it could breathe oxygen levels at around 30 percent or thereabouts. Today, the oxygen content of the air around most urban and suburban areas is around 19 to 21 percent and decreasing. In larger cities, the oxygen content can be as low as 12 or 15 percent, making it incredibly difficult for the body to get enough oxygen for healthy living. Some think these decreasing oxygen levels could be behind the increasing rate of cancer worldwide. Tailpipe and smokestack pollution may therefore go hand in hand with both climate change and the relatively new phenomenon of cancer.

The best method for carbon 'capture' is, not surprisingly, planting trees. However, what is surprising is the amount of land in the world that has been deforested yet remains unused for anything else. In other words, people throughout history have stripped areas of land for the wood and then left it bare.

Reforestation schemes seek to recreate natural woodland areas, restoring them to their natural and original beauty. 'Lazy reforestation' programmes seek to plant quick-growing species and usually only one or two different types. This can result in monocultures and a stifling of natural ecosystems. In parts of the world that are arid or tropical, once trees are felled the land can become too dry for new forest to take over again. Also, over-grazing and over-harvesting of forest resources may be occurring, and both can lead to desertification and a loss of nutrient-rich topsoil. Lazy reforestation in these areas can be very unsuccessful, and waste a great deal of time and resources. A successful reforestation program must consider both the native species required and employ a healthy mix of them all.

Reforestation is nothing but a good thing, as long as nobody is forced off his or her land unwillingly, or other types of ecosystems are cleared to make way. If these conditions are met a new forest can bring many benefits: shelter for multiple species, reintroduction of threatened species into an area, use of disused land, and massive carbon sinks.

But reforestation, some experts say, isn't the best solution. Some point out that the scale of the worldwide scheme would be enormous, and the speed would also have to be extremely rapid. For a start, the rainforests, which currently recycle 50 percent of the carbon dioxide and oxygen in the world, exist within national boundaries, therefore giving a fraction of the planet's population the right to do as they please with them. Secondly, reforesting the rainforest is difficult because, once the trees are cut down, the soils quickly lose fertility. Temperate forests grow very slowly.

In light of this, many scientists have been working on artificial methods for removing carbon dioxide from the atmosphere, and some are hitting the headlines. One of them is what is known as the *Synthetic Tree*. Geo-physicist Professor Klaus Lackner has designed a device that basically mimics the process of sucking CO_2 out of the air that real trees use. Synthetic trees would not use photosynthesis however, and this would mean that its leaves can be packed more tightly and this, Lackner claims, it what makes them much better than real trees at removing CO_2.

Lackner's trees would use a solution of sodium hydroxide, which absorbs carbon dioxide from the air when it comes into contact. The result is sodium carbonate. This could then be piped away and the CO_2 recovered again for final storage, which Lancker suggest should be deep underground at the bottom of the sea, where rock is porous. Using the same technology of a marine oilrig, the gas could be stored thousands of metres under the ocean bedrock and become locked away, probably for millions of years.

However, it would take many thousands of synthetic trees, each several storeys high, to compensate for the emissions of CO_2 currently being pumped out. This would take up a large area of land, though Lackner does point out that the 'trees' could be put anywhere on earth and still have the same effect, so they could be placed at sea or in barren deserts. This may help ease the vision of huge construction projects in the countryside, rivalling the apparent eyesores of wind turbines, but it does nothing to tackle the problem of cost. Still, each 'tree' should be able to remove 90,000 tonnes of CO_2 from the atmosphere every year, the equivalent emissions of 20,000 cars, making the idea a credible one.[iii]

A second plan seeks to use nature's photosynthesising power to remove CO_2 from the atmosphere, only without using trees. Instead, some propose a method known as *Phytoplankton Fertilization*, an idea that attempts to take advantage of one of the world's largest destinations of carbon dioxide.

Phytoplankton is microscopic plants that live in the oceans, and not only provides a basis for marine food chains, but they also absorb carbon dioxide from the water, and therefore the atmosphere. One could swim through phytoplankton and never know about it, but from space it is possible to see the huge green clouds just below the surface of the oceans. They gather around coastal waters and river estuaries because they are fertilized by nutrients in soil that is naturally washed across the land and down into the sea. Like land-based vegetation, plankton uses photosynthesis to grow and self-sustain - and hence remove vast quantities of CO_2 from the water like forests do from the surrounding air. Additionally, it also means that they emit vast quantities of oxygen. Like land vegetation, the death of plankton doesn't suddenly release CO_2 back into the atmosphere but actually takes it to the seabed instead, where it will rest for tens of thousands of years.

In 1995, a team of oceanographers headed out to an area 400km south west of the Galapagos Islands known as the 'desolate zone' because it contains so little plankton. The team speculated that the desolate zone was missing iron, one of the nutrients usually washed from soil to the sea, and therefore making the zone a desert for plankton. Dr John Martin of Moss Landing Marine Lab in California developed a method of fertilizing plankton with an iron sulphate solution. Sadly, Dr Martin died of cancer before the ocean test could be put into practice, but the team that went out proved convincingly that this method dramatically increased the plankton bloom, witnessing it multiply up to 80 times faster than ordinary growth. Half a tonne of iron was added to the ocean water and the ocean turned green with plankton. By the time the experiment was finished, the scientists estimated that the plankton had absorbed an extra 7000 tonnes of CO_2, the equivalent of 2000 fully grown redwood trees.

But iron is quite a common substance in 80 percent of the world's ocean waters, and so dumping more of it into the sea would not produce very dramatic plankton blooming. Instead Professor Ian Jones, from the University of Sydney, plans to use the nitrogen rich substance urea – a major component of urine. His plan is to turn granulated urea (bitter tasting pellets) into liquid form, and then pipe it out to parts of the ocean that severely lack plankton.

With a booming human (and cattle) population on Earth, urea would be a very cheap and common substance to get hold of, especially in Rich Northern countries where sewage systems are already in place and of high standard (for the record, it doesn't help fertilization just to go and piddle in the ocean – urine and urea are not the same!).

Then again, what effect will adding gallons and gallons of fertilizer have on marine ecosystems? In the past, whenever large quantities of nutrients have been added to bodies of water, including the sea, there is a spurt of bacteria growth, which depletes water of oxygen and can kill fish and other organisms. However, Professor Jones understands this problem well: "we are not doing it where there is lots of productivity – we are doing it in the desert region of the ocean, where very few marine organisms live already."[iv]

Such a scheme would also need round the clock fertilization taking place, since plankton is quick to return to its natural levels once nutrients are no longer coming in excessively. But this may also be a blessing in disguise. If we suddenly found there to be an unwanted problem with the plan, perhaps something unforeseen, then we could turn off the tap and the oceans would return to normal very quickly, as if nothing had happened.

Two more big schemes are being proposed to keep climate change at bay, but unlike the first three, neither seeks to remove carbon dioxide from the air. How do they plan to do it? By blocking out the sun's radiation. The first plan proposes creating clouds and plumping up existing ones. Clouds perform a major role in the planet's natural regulation of climate, blocking out solar radiation and cooling the planet slightly. Now some scientists want to make this effect happen on a much larger scale – *Cloud Seeding*.

Surprisingly the idea of fertilizing clouds goes all the way back to 1946, when scientists found that firing rockets full of silver iodide particles into rain-bearing clouds made it rain. Professors John Latham and Stephen Salter don't want it to rain more, but playing on the idea they are now in the process of developing a scheme for increasing cloud reflectivity. Low-level clouds, which cover about a quarter of the world's oceans, act like giant mirrors, sending large quantities of solar radiation back into space; Latham and Salter want to increase how much light these clouds can reflect, and therefore slowing global warming. They calculated that an increased reflectivity of clouds by just three percent would cool the planet enough to compensate for all of human-induced global warming.

The scientists plan to create clouds out at sea, spraying seawater droplets continuously high into the air at a rate of 50 cubic metres per second. Once up there the salty droplets would attract fresh water droplets and build cloud cover. To do this would require a lot of propulsion, as well as a floating device to allow the clouds to be built up over wide areas at sea. The plan is for thousands of vessels to be deployed, powered by Flettner rotors - spinning vertical cylinders mounted on the vessel, out of the top of which the seawater would be ejected into the air. The power would come from turbines under water but instead of being powered by something the rotors would be dragged through the water by the vessel and actually be the source of power. This power should cover all the electrical needs without use of electricity from batteries or even fuel cells.

Though this scheme utilizes probably the only substance we seem to have far too much of on this planet – sea water – the ecological effects of creating clouds is still unknown. It might interfere with current rainfall patterns, and possibly worsen the rainfall levels in countries already suffering from drought. Then again, like the fertilization of plankton, if this scheme has negative consequences it can be switched off, with the world returning to normal quickly.

And if firing sea water high into the sky and piling up clouds doesn't sound extreme enough, try the second of the schemes: *The Sulphur Blanket.* Like Cloud Seeding, the Sulphur Blanket leaves CO_2 well alone and instead aims to tackle the incoming solar radiation.

The idea goes back to volcanoes and tectonic climate change. In 1991, Mount Pinatubo in South East Asia blew its top and ejected about ten million tonnes of red-hot sulphur into the stratosphere. Scientists calculated how much sulphur had been injected and then watched as it spread around the world. After about a year the sulphur had spread quite evenly around the entire stratosphere and for two years after the eruption the world's average temperature dropped by 0.6°C.

Professor Paul Crutzen was one of those scientists watching Pinatubo's effects and proposes duplicating it to create a giant sulphur screen against incoming radiation. To do this, he says, would take hundreds of rockets filled with sulphur, altogether sending a million tonnes of the stuff up there. But sulphur is sulphur, and down here on the planet's surface it has already established for itself a rich history of environmental degradation and human health problems. Even to non-scientists, the idea of creating sulphur clouds all over the world seems dangerous. Crutzen argues that the sulphur will be injected into the stratosphere, the second layer of the atmosphere – therefore it will be too high to affect the surface.

But problems remain; more sulphur in the atmosphere could increase both acid rain and the size of the ozone hole. Furthermore, there remains the yet unproven story of what happens to all that sulphur once we no longer need it in the atmosphere: how long will it take to clear away? But as Crutzen says, losing a bit of ozone or facing unknown consequences may be all part of a necessary evil now that climate change is reaching a critical stage. Long gone is the time when we could worry about such luxuries. This argument pops up now and again when it comes to the undesirable side of climate change – as it did with nuclear power and onshore turbines – emphasising again and again how late in the game we have waited to solve the climate crisis.

So which of the grand schemes stands up to reality? The latter two are unproven to really work, and may cost billions of dollars to get past development stage. In the case of the Sulphur Blanket, playing around with the atmosphere might even be very dangerous. For the Synthetic Trees the downfall might come at the cost and the time for development. To build so many large devices and install them around the world will be very costly, and there is no long-term, guaranteed supply of sodium hydroxide, the key component. Which leaves us with fertilization of phytoplankton; this idea seems to have the greatest merit out of all the schemes aimed at buying us time. The fact that urea is so readily available, and that plankton is so good at removing CO_2 from the atmosphere, helps it enormously. Its apparent low cost is also a large plus, and means it could be applied fairly soon.

Unfortunately, none of the schemes tackles what is the root cause of climate change, and all the designers admit that they are reluctant advocates of their plans. In an ideal world, humanity would learn to live without such a big demand for energy, green or not, and learn to ditch fossil fuels. We don't live in that ideal world, and fears about the economic costs of switching off fossil fuels and switching on alternative sources seem to outweigh anyone's fears about the safety of life on Earth. That is why we will probably need these schemes; no matter how crazy they may seem to us at first.

In conclusion, our booming energy demand is driving the Earth into catastrophe. We still get around eight tenths of the world's energy from fossil fuels when cheap and effective renewable resources have existed for decades. Future power production lingers on the horizon just out of reach, and we may need carbon capture or cloud creation schemes to buy us time to get there.

The bulk of the problem perhaps lies with the private ownership of energy. The solution may be found with decentralised energy sources (each home produces its own energy), simultaneously making use of the parts of the world that are best for certain types of energy production (such as solar heating in deserts) and then sharing all this on a continental or even global power grid. Perhaps our energy needs to transcend national borders and private ownership.

Sweden is one shining light in all the fossil fuel murk. The country, already one of the most environmentally friendly, is aiming to be *completely* free of fossil fuels by 2020. That is an incredible ambition. It already had a head start, owing largely to its green politicians, with 26 percent of all its energy coming from renewable sources (the EU average is 6 percent). It gets large amounts of its energy from nuclear and hydroelectrical sources, but its renewable sector is ready for a big boost. It also plans to bring in alternative fuels for all vehicles, phasing out petrol as soon as possible, offering grants to public sector services to convert from oil use, and green tax breaks for homeowners. Sweden is a cold country, but if they can keep themselves warm without heating up the planet in the process, then there's no reason why the rest of the world can't follow.

---

[i] 'Where the dream of harnessing the sun's power could come true,' James Randerson, *Guardian Online* website, May 24th 2006. http://www.guardian.co.uk/print/0,,329487778-117780,00.html, Accessed 24th May 2006
[ii] Smith D. (2003). *The State of the World Atlas*. Earthscan Publications Ltd, London. Pg 31
[iii] 'Artificial Trees: a green solution?' BBC News [online] February 20th 2007. http://news.bbc.co..uk/go/pr/fr/-/1/hi/programmes/6374967.stm, Accessed 20th Feb 2007.
[iv] 'Multiplying the ocean's CO2 guzzlers,' BBC News [online] February 19th 2007. http://news.bbc.co.uk/go/pr/fr/-/1/hi/programmes/6369401.stm, Accessed 20th February 2007.

Chapter 12 – Begin by Defending Yourself

The past few years have seen a notable turn around in the media regarding climate change, at least in Europe. A journalist who questions human-induced climate change is now going to fall into the minority rather than the majority, as before. This is a welcome change from the Age of Denial we seemed to fall into during the 1980s and 1990s, and hopefully signals the start of the Age of Awakening. But the media is cautious when it comes to changing its mind. For years, mainstream media had been selling the climate change debate as a 50-50 duel; it might be, it might not... the jury's still out. Even now, while the mainstream begins to embrace the possibility of what could happen, we still see the tiny minority that are the 'climate sceptics' given an even representation in the news.

And then there's the stubborn nature of human laziness. Though individuals may be motivated to get off the couch and do something about climate change, the majority of people in the Rich North still see climate change as they saw it years ago: an issue for the future, an unproven theory, an issue for governments or NGOs to sort out. So even with a slow u-turn of the mainstream media in favour of reports on climate change, it takes time for the public to accept it. That's why when you talk to friends and family about climate change someone is likely to pop up and mutter something about 'the other side to the story'.

Why are we in denial?

Scientists don't help:

Science is, by rule, 'small c' conservative. It is very rare that any scientists will go beyond the data they have and make a firm assertion about its consequences. This is only an obvious and logical step in science - after all, if scientists let their imagination loose at any given whim, we could be in a real state of mass hysteria every time a scientific story hit the news. It is part of a scientist's training to be sceptical. This is why pharmaceuticals have to be tested over and over, again and again, and why any sort of study should include hundreds, if not thousands, of individual data if it is going to be credible.

When scientists see glaciers melting, or temperatures rising, or sea level rising, or greenhouse gas emissions rising, they have to remain calm and collected. Scientists on television always appear to be the calmest individuals in the world. For a scientist to come out and declare radical change is to risk a career stacking shelves in a supermarket. Inside they might be alarmed, but the eyes and frowns of the scientific community are ever-watching.

We are still ignorant of the Earth:

Earth is fantastically complex, and there are so many different climate influences and feedback loops that we are learning more and more with every passing day. It would be hard enough predicting the future climate if the system were simple, but it's not, and with every discovery, adjustments have to be made all. Pick up a textbook on climate from a few years ago and it will seem a little outdated – already; a climate book from twenty years ago isn't really worth the paper it's printed on.

We are not used to things happening so quickly:

In our everyday lives we encounter gradual change. With global issues or demographics this is always the case, and as a social animal we are only used to changing circumstances if they take a long time to happen. Outwardly, signs of the aging process are usually too small to notice. Our wages creep up with inflation or with promotion. Our nine to five job-orientated societies are monotonous, and special occasions only happen on special occasions. Our brains simply can't get used to exponential changes or rapid shifts in the world around us. We've never been faced with a runaway situation before, where everything spirals out of control - and if we have, it has only been transient and localised.

Therefore, it is difficult for people to accept that the Earth in ten years' time will be very different from that of ten years ago. Hence why we have so many problems with starvation and poverty in the world today... we simply weren't prepared for the population explosion of the twentieth century.

We are not used to global issues and global threats:

Human history has never known a global crisis before. Both World Wars were huge and devastating events, for all involved directly and indirectly, but the very existence of our species was never threatened. Climate change has the potential to ruin every human life on the planet, whether you are a Polish Jew, American Christian, tribal Amazonian, Tibetan Buddhist, radical fascist, Stalinist, Quaker, vegetarian, plumber or housewife... or whatever. *Everyone* will be affected unless radical change happens fast – *that* is new, and that we are not prepared to deal with.

The media is mostly private:

Most media outlets are privately owned. Worldwide, the vast array of media outlets and their numerous competitors are owned by only a handful of different companies. Due to the nature of business, nearly all of these parent companies have conservative interests or agendas.

There was a powerful and ferocious denial campaign in the 1990s that won large successes in turning climate change back into a theory. One way in which this happened was to get scientists onto the side of those who have interests in maintaining the fossil fuel status quo. Oil companies are particularly good at making sure they have some 'experts' in their corner, ready to come out swinging whenever a 'serious scientific debate' is called for. In an effort to remain unbiased, the media often gives equal representation to such people as they do to serious climatologists. The media appears fair, the debate seems open and the public is none the wiser.

Answering those Frequently Asked Questions

Remark: Though greenhouse gas concentrations have increased noticeably since 1900 there has been no significant or equal rise in global temperature over this time.

Response: Greenhouse gases build up in the atmosphere. At low levels the majority of these greenhouse gases are absorbed by the planet (such as oceans and forests) and so the overall gain is quite slow. Additionally, there is the natural variation of the Earth's changing temperature that has to be taken into account, and if there is natural warming then the artificial warming will be disguised or hidden within. To deny that greenhouse gas rises have contributed to temperature rises is to deny the validity of the term 'greenhouse gas.'

Secondly, there is a delay period that has to be taken into account. Just because greenhouse gas emissions and temperature increases don't match perfectly, doesn't mean there is no link. The climate will not match the changes in atmospheric greenhouse gases exactly and some short-term changes may seem to contradict themselves. Since around 2000, climatologists have begun to pick out the signal of artificial climate change from the natural variation, showing that the build up of greenhouse gases has finally spilled over the natural change. They have begun to show that the Earth is warming thanks to human pollution, regardless of the Earth's natural changes.

Remark: the climate changes naturally anyway, given time. The change we are seeing now is only part of a natural cycle.

Response: Every single change in climate that we have observed from ice cores, tree rings, lake sediments and other natural features, show that all took place with natural variations in climate that are much smaller than the ones we are causing today. Today's carbon dioxide levels are the highest they have ever been in the last 650,000 years, putting us outside the natural variation. If the cause is unnatural then the consequence always will be too.

This simple thought experiment helps: if a fire is fuelled by wood and its flames flicker higher at some times and lower at others, what will happen if you add more wood? The answer is the fire will grow, and the flames will flicker higher than before and not as low. Suggesting that adding thousands of millions of tonnes of CO_2 into the atmosphere isn't going to cause greenhouse warming is like saying that adding fuel to a fire won't make it bigger.

The idea that humans aren't causing climate change is utterly unscientific. Sceptics say that changes in the planet's orbit and not the greenhouse effect are causing the warming. This is a simple way of getting humans off the hook. Consider this: recent studies since 2001, including one by NASA, has found that less infrared radiation is escaping into space as the years go by, and that the planet is now absorbing one watt per square metre of its surface more than it releases.[i]

Remark: Carbon dioxide is a poor greenhouse gas. The real greenhouse effect comes from water vapour.

Response: Unfortunately this is only a half-truth, and a deceptive one at that. Although water vapour is indeed the major and most potent greenhouse gas, it only begins to affect climate as a positive feedback, in response to other changes beforehand. This feedback works only when the relationship between air temperature and pressure is changed to a certain degree; therefore it only responds to our already changing the air temperature. It is thought that water vapour cannot change in a completely independent manner and become the initial cause of climate change.

Remark: Even scientists do not agree that humans are causing climate change.

Response: Wrong. Very few scientists remain that deny the link between human emissions and climate change – just as there are still a few people out there who deny that the Earth orbits the Sun. In fact, scientific consensus is rarely as strong about issues as it is about humans causing climate change. The widely held scientific belief is clear – climate is getting warmer, it is caused by humans, and it will only get worse if greenhouse gas emissions continue to rise.

Remark: There is no link between temperature increases and increased CO_2. The rapid warming between 1920 and 1955 occurred during a time when *greenhouse gas emissions were rather low, and as greenhouse gas emissions increased between 1950 and 1970 the global temperature actually cooled.*

Response: Again this brings us back to the complexity of the climate system. On any time-scale in Earth's history, there have been periods that have contradicted the overall, long-term change. The gradual cooling of the last 55 million years has been interrupted several times by periods of warming, caused by several changes in the climate components. Still, the gradual cooling continued. When the northern hemisphere began to warm, and bring about an end to the last glacial maximum, the warming trend was interrupted by a period of cooling (Younger Dryas). Still, the global warming continued and the major ice sheets melted (except Greenland).[ii]

Blips here and there do nothing to undermine the overall will of the planet to change. The cooling of the 1960s and 1970s fails to prove anything because it occurs on top of a global warming trend beginning around 1900 and following on into the present.

Remark: How can you talk about 'global' warming when some places on earth have not changed at all or have in fact become cooler in recent years?

Response: Earth's climate is incredibly complex. It is silly to presume that every part of the climate will change in the same way or to the same degree. In some places, the cooling effect of sulphate clouds has masked the overall warming of the surface, and sometimes caused it to cool. In other places, weather patterns appear to have become cooler as the climate changes. This is all a response to the overall warming. Global warming is better described as 'climate change' because it takes into account the unnatural cooling caused by humans, such as the potential freezing of Northern Europe if the North Atlantic Ocean Conveyor shuts down and stops transporting heat from the Gulf of Mexico.

The majority of places on earth are warming all the time. In some places the climate is already dry enough, so further aridity or fewer monsoon rains will send populations into panic. We may look out of our window and see no change to support climate change, and feel comforted by the predictability of our weather, but if we lived around the Arctic Circle we would understand climate change was a real and devastating threat, which is building year upon year. The build-up is undeniable: the unprecedented weather of 2005, the increasing global temperatures, larger deserts, milder winters, climate records being broken again and again with increased frequency… it all accumulates to prove that the globe is warming.

Remark: Global warming is good. It will give us warmer and longer summers, milder winters and make plants grow quickly.

Response: Particularly in places like the UK this argument is strong, because people are carried away by thoughts of Mediterranean summers, vineyards along the South coast and beach holidays. When one lives in a cold place it is hard to think of any kind of warming as a bad thing.

The problem comes when one starts to think outside the local area. They might appreciate a little warmth in Chicago, but what about the sticky subway commuters of New York, the farmers in the central US, or the vineyards of California? What about the oceans, which are warming and killing off vast areas of coral reef – crucial sources of food and shelter for creatures at every stage of the ocean food chain? What about the sea levels, which continue to rise and will soon threaten coastal cities, turning millions of people in refugees? What about increased drought in the Sahel and Asia, and more severe flooding? What about the more intense storms, the spread of disease, the lack of food and clean water, and the mass extinction of species sure to follow?

Remark: Why do you care so much about animals and trees and the air when there are so many people out there starving to death as we speak? Do you not agree that combating poverty is a lot more important than stopping some temperature rises?

Response: Again this is a popular argument against action on climate change, usually initiated by people who have a real experience of poverty. For some, the climate change problem should be secondary to the problem of human suffering on Earth, both in the Rich North and the Poor South. Around the world today, about a third of people live on or below the poverty line, millions more work for big corporations for only enough pay to feed themselves. Despite popular opinion, the gap between Rich and Poor in this world is widening. More people live with malnutrition, homelessness and disease than ever before in human history. Something has to be done.

It is hard to say anything at all that seems to argue against feeding the hungry or helping the poor, because poverty is an issue so close to us, affecting men, women and children, in Sudan or Seattle, Palestine or Paris. But climate change is a ticking time-bomb – when it explodes the poverty problem will explode with it. Africa already has a problem with food shortages, poor crop yields and malnutrition; rising temperatures bring drier soils, fewer rains, and the threat of flooding that washes away soil nutrients. In Asia, the problem is also huge, as is the threat of global warming. Literally thousands of millions of people could die as a result of global warming, far more than today as a result of poverty.

In all parts of the world there will begin a scramble for resources. What little arable land and fresh water supplies remain will be contested over. Prices will rise, medicine will be needed and the only people to afford all this will be the minority of the rich. The first victims of global warming will be the poor, and many of them will pay with their lives.

---

[i] Pearce, F. (2006). *The Last Generation*. Transworld Publishers, London. Pg 29.

[ii] Ruddiman, W. F. (2001). *Earth's Climate: Past and Future*. W. H. Freeman & Co., New York. Pg 421.

Chapter 13 – Finish by defending the planet

'People are often very open-minded about new things – as long as they're exactly like the old ones.' – Charles F. Kettering, 1876-1958, American electrical engineer.

When all is said and done what really matters is that we learn our lessons. Understanding climate change at scientific and historical levels is of no use to us if we are going to carry on as we always have done. As the great Sam Cooke once said, 'change gonna come' – it is up to each and every one of us to decide if *we* are going to change, or if the climate is going to change.

Personal Change

Though climate change can only be stopped with a huge international effort to change our energy supply, the easiest things to do to make the world a better place usually start right at home. Almost every single person in the Rich North lives a lifestyle that is unsustainable and exerts a hefty burden on the poor planet. In fact, some of these changes are so ludicrously simple to make that it really is surprising that we fail to do them now. Other changes take a little effort, but even these can be made so simple and fun that you don't have to think about them. Some seem like hard work on paper, but when you start doing them it all becomes so easy. Every change listed below, no matter how small, is needed to put the brakes on climate change. Perhaps the best thing about saving the planet is that it can also save you a lot of money.

1. Start at home

Don't use standby. A television produced before 2005 may use 50 per cent of the energy on standby as it would if it were fully on. Since there are roughly 250 million working television sets in the United States alone, the saving on energy could be enormous around the world.[i] Ignoring the standby function and turning devices off completely saves around 0.02 tonnes of CO_2

per year, per device.[ii] Other appliances consume energy almost as much as television sets when they are "off" – DVD players, mobile telephone chargers, stereos, computers, or any other appliance that has internal memory, permanent display, remote control or charger. Such appliances need only to be unplugged to save this energy.

Get energy efficient light bulbs. Lighting is perhaps the most common use of electrical energy in the world, and if we want to save energy we should definitely start here. Old-fashioned incandescent light bulbs are extremely inefficient with the electricity that goes through them. Nine-tenths of the entire energy passing into the bulb is lost in the form of heat, as you may know if you've ever touched one in use. This means that they are ten per cent energy efficient. Energy efficient light bulbs last up to 10,000 hours – ten times longer than their old-fashioned cousins – and consume up to 75 per cent less energy. Turning off lights whenever they are not needed typically saves 0.3 tonnes of CO_2 per household per year.

Buy energy efficient appliances. Everyday household appliances, such as refrigerators, water heaters, washing machines and air conditioners, all use incredible amounts of energy. Appliance manufacturers are always improving the efficiency of their products for a more energy-conscious market, and often such products are graded on a scale of energy efficiency at the point of sale (in the UK this scale is from A to E with A-grade products being the most efficient). Grading products makes it easier for the consumer to make purchases with energy consumption in mind.

Try to fix appliances before throwing them away. What we must keep in mind before we go out to buy new products is whether we actually need them. In case of home appliances, it would be far better for the environment to keep using old refrigerators and washing machines etc. until they break down, rather than throwing them away when they still work in favour of newer models – even if they are more efficient.

When appliances do stop working, our society is far too eager to throw them away and get new. This urge is also driven by the desire to be fashionable in the home, to refresh and update the house. However, many times a broken appliance will only need a small repair to get up and running again. Far too often in our society our landfill sites are filled with toasters that still toast, refrigerators that still refrigerate and microwaves that still microwave. It's a waste of resources and energy.

Insulate your house. Every building ever built loses heating somewhere. Heating and cooling can account for almost half of a building's energy use, which is astounding when you think about all the electrical devices we have nowadays. Of course, generating heating for our homes and offices is an essential part of life in countries prone to cold weather. But buildings are poor at keeping heat in, with the major outlets being windows, rooftops and around doors. As a result, the majority of the energy we use warming or cooling ourselves in the home is lost. In other words, you actually spend energy and money heating/cooling the rest of the world rather than just your living room.

There are several things you can do to prevent energy loss in the home. Putting at least 15cm of insulation in the loft will save 20 per cent in heating costs.[iii] Insulating the hot water heater and water pipes also allows less energy to be spent warming the surrounding air. Windows can be double or triple glazed, trapping more warmth inside. Draughts around doors and windows can be sealed easily enough, and can make big differences. Insulating all roofs, floor pipes and tanks could save emissions by as much as 1,260kg per year.

Heat your house efficiently. Lowering the thermostat temperature by just one degree can save a lot of money. To combat the cold, often the best solution is to wear warmer clothes.

Conserve water. Traditionally (pre-1970), most people saved energy and money because they had no real choice about it. Imagine Victorian Britons leaving the fire blazing when they went out for the day, or throwing things away that still worked, just because they fancied a change. It just wouldn't happen, because people once knew the value of saving money. Heating water is one of the biggest energy uses in the home and also one of the most wasted. Some water heaters are set to such ridiculously high temperatures that water comes out of the tap almost as steam and people use dishwashers and washing machines on half-load or almost empty. Having showers instead of baths can save water and energy.

Switch to green power. Of course, we could zap up all the energy we want and live a lazy and wasteful existence until our heart's content, if only our energy came from green sources. Unfortunately, so little worldwide energy is produced by renewable sources that we cannot afford to be so wasteful. Thankfully, it is now easier to switch to green energy sources than ever before, and often it requires one click of the mouse or one telephone call. The more demand for renewable energy, the greater the proportion of energy production renewable energy can attain.

If every home produced its own energy the savings on energy consumption would be greater and the cost less. In many cases, such a solar panels, the government may give you a subsidy to have them installed, cutting the cost right down. Excess energy can be sold back to the energy companies/government and delivered through the national grid. If every home in the world produced its own energy, using the best of the environment it had around it, then there would only need to be a handful of power stations built to run other energy uses, and they could all be renewable.

2. Out and about

Avoid the car, love the bike. The automobile is fast becoming the bane of humanity, zipping people to and fro in a cloud of greenhouse gas emissions and poisonous carbon monoxide. In the US, almost one-third of CO_2 comes from cars, lorries, trucks, airplanes and other vehicles that are used for transport or for delivering goods.

It all comes down to convenience. We can get to almost anywhere within a reasonable range using public transport, cycling or walking, but the average person would rather get from point A to point B with minimum of effort. Walking is the best thing to do because it requires nothing but you and your legs. But in this hectic society walking becomes impractical. If it is possible, walking can also benefit your health greatly. Sometimes cyclists get to their destination before drivers, especially when everyone else is trapped in traffic jams or at road-works. If walking or cycling are out of the question then consider public transport. Buses and trains may use more fuel than a car, but one bus of three people saves more than three cars with one person in each. The New York subway system is one of the most popular mass transit systems in the world, because it is cheap, quick and there are plenty of trains.

For some there is no place like the car, even if it costs more on petrol than the price of the train ticket. If public transport can't be stomached, carpooling can cut the journey of five or six different cars down to just one, and helps you make some friends on the way.

Can your next car be energy efficient? With each new generation of cars that are produced the fuel efficiency increases. But saving emissions means more than just updating your car every now and again for a newer model. Hybrid cars are powered by both petrol and electricity - making it one of the most energy efficient automobiles available to buy, regardless of its make and model. These cars are still quite rare simply because the demand for them is too low; if more went out and bought a hybrid the price would fall.

See http://www.eta.co.uk for environmental performance of new vehicles.

For hybrids see the US website http://www.hybridcars.com, and to learn more about fuel cells see http://www.fueleconomy.gov/feg/fuelcell.shtml.

3. Life

Buy fewer things. Since about 1950, there has been a new driver that has kept energy needs soaring in the developed world, and which also forces population, energy and deforestation increases in the underdeveloped world: consumerism. As the rich get richer, thanks to consumerism, the poverty gap grows; poor populations boom and demands keep growing in the Poor South. It becomes an endless cycle, with the poor forever trying to catch up with the rich. Our incessant demand for more and more consumer items, that we simply don't need, feeds the cycle that causes global warming. Businesses create products and services that we really don't need, and as they do so they burn fossil fuel energy in factories and deliveries.

Even if consumerism was fuelled by renewable energy, extracting resources exerts a huge pressure on the natural world. Therefore, perhaps the best thing we could possibly do in our every day lives is to stop buying things! Next time you are about to buy something ask yourself if you really need it. Can you live without it? If not, can you borrow it from someone else?

An excellent site is http://www.newdream.org.

Buy local. Over the last couple of decades it has become more evident that the rise of big retail companies – such as supermarkets - has begun to squeeze out small, local businesses. The concentration of shops in town centres, and away from villages and local communities, has been very negative for smaller businesses. Large companies now import products from places on the other side of the world to sell them in the domestic market. Not a single ship or aeroplane used in commercial shipping is fossil fuel free, and the further that ship or plane has to travel the more fossil fuels are burnt.

It is an ugly consequence of our world economic system that it is cheaper for companies to produce items in poor countries for slave wages and then ship them back to the rich countries than to simply produce them in the rich countries. It costs our planet dearly. Nearly all products should have country of origin stamped on. This is where the idea of food miles comes from. Even better, grow your own food, if you can, in allotments or your own garden. This can also prove a good workout and a hobby if nothing else.

See http://www.farmersmarket.net and http://www.soilassociation.org for information on local food and markets.

Also try http://www.sustainweb.org to push for better food, and http://www.bbc.co.uk/food/food_matters/foodmiles.shtml.

Re-usable bags. Plastic bags really are the blight of the natural world. They get everywhere, from landfill sites to remote mountain forests – scientists have even found them floating around Antarctica, about as far away from consumer civilisation as you can get! Apart from killing animals willy nilly, it takes a lot of energy to create the billions and billions of plastic bags that are produced annually. It also justifies the extraction of oil. Look for any reusable bags that might be on sale at the shops, or take your own along. Often these bags are very large too, so a whole weeks food shopping can be done using only three or four reusable bags.

Don't waste paper. The production of paper is one of the most energy intensive industries and it causes indescribable amounts of pollution – and then there is the whole deforestation issue. Of course, some might argue that as long as paper comes from a forest specially grown to produce paper then it is alright: carbon neutral. But this is wrong. Before worldwide deforestation intensified around 1700, and ever since, many parts of the world used to be covered in forest. In places where paper forests are planted, there used to be forest anyway, so there has been no gain or even a break even of forest. Furthermore, paper forests are usually fast growing species of tree that are not native. Reduce paper consumption and new forests can become carbon sinks.

Reuse, repair, and recycle. Reducing purchases is the best way to reduce waste because it prevents more materials being used in the first place. Reusing purchases acts as the second defence against waste. Most items can be used more than once, even for a different job altogether. If it remains unneeded, someone else may want it instead. If items cannot be reused because they are broken, they may be fixable. For nearly all products, repair consumes less energy and materials than producing a new one, and often a repair can be as easy as fiddling with a few knobs or tweaking a few wires.

Recycling should always be low down the list of actions to take with unwanted products. Recycling can be a very energy intensive process, involving large recycling plants and a collection service no doubt running on fossil fuels. But, after reuse and repair, recycling is the best option; it puts materials back into the production cycle, reducing the need to acquire raw materials.

Compost food and garden waste. When we send organic materials into landfills we are advocating the production of methane - the most potent greenhouse gas. Methane is produced when decomposition occurs without vital oxygen. Alternatively, composted organic waste adds precious nutrients to soils and can prove excellent for gardeners and farmers in the growth of healthy and nutrient-rich foods.

Eat less meat. The meat industry is one of the biggest and yet unnoticed contributors to climate change. It takes roughly 1000 litres of water to produce one kilogram of wheat crop, and 100,000 litres to produce one kilogram of beef. A hundred times more water demands a lot more irrigation for farmers, which then goes on to compound the fact that more forests are being cut down around the world for livestock grazing. It also takes more energy per-unit to produce and transport meat than it does for an equal amount of plants, such as wheat and vegetables. Additionally, the demand for fishing and bush-meat both contribute to extinction.

See http://www.epa.gov/methane/rlep/faq.html, and http://www.earthsave.org/globalwarming.htm.

Change the system

Ultimately all this light bulb changing and home insulation isn't going to be very effective if the powers-that-be continue business as usual. Why do companies produce televisions with standby buttons in the first place? Why does the government push the public to save energy when it could reduce emissions more effectively by switching off fossil fuels?

1. Inspiring change in others

Become a climate teacher. This doesn't have to be hard or time consuming. Just let your friends and family know how they can cut their own energy uses around the home and with transport. Get a public meeting going where you live and explain to your neighbours what the problem is, what causes it, and how it can be stopped. The problem with climate change is that it requires us to act in unity before we can stop it; there are no solo heroes involved. Having all this knowledge is useless unless you share it with other people and encourage them to change their lifestyles.

Demand change at work. Requesting that your company does its part to tackle climate change can be fairly simple. Switching off computers and office lights at night, when they are not in use, obviously saves energy, as does using less paper. Video-conferencing or making a telephone call may be a better alternative to travelling to meetings – especially one that requires getting aboard a plane. Remind your boss that companies with good green credentials have a largely improved image amongst the public and prospective clients.

2. Demand change

Support environmental groups. If possible, regular donations to decent environmental organisations can make a real difference. Organisations like Greenpeace and Friends of the Earth have been fighting local and global environmental catastrophes for decades now and have the expertise and infrastructure to take it to politicians and business leaders with real tenacity and commitment. Groups like these have been trying to warn the world about climate change since the issue first entered the public consciousness.

Vote Green. In decades passed, Green Parties around the world have been seen as irrelevant and not to be taken seriously. Without the big money donations that the mainstream parties receive, Green Parties find it difficult to compete even at a local level, let alone nationally. But, for the next hundred years, green issues are going to be the number one issues of the times, and who better to deal with them than people already keyed in to the causes and solutions? Certainly not mainstream parties who have always ignored green issues.

The UK Green Party has proved it is able to make real changes where mainstream political parties have only produced a mirage of promises. In the little Yorkshire area of Kirklees, the local Green councillor has managed to push for better social housing, with the town now having five per cent of the UK's rooftop solar-photovoltaic panels. Improving insulation has helped to reduce bills and poverty, as well as emissions. Council houses in Kirklees generated fifteen per cent of their electricity and sixty per cent of their hot water needs by 2006.

Many people perceive a Green vote as a wasted vote, but even if the party don't win this is still a false assumption. The more councillors a party have on a council the more the balance of power switches in their favour. Repeatedly throughout this Labour government, Labour councillors had to seek help from Green councillors in order to get their own ideas passed. In return, the Green councillors can make their own demands.

Demand Green energy. Renewable energy only accounts for 0.4 per cent of all energy production on Earth, meaning that climate change is unavoidable if we are to live the lifestyles we currently live. Fossil fuel power stations devour millions of tonnes of fossil fuels every day, and nuclear power stations use more energy in construction and deconstruction than is worth the greenhouse gas emissions they save. On the whole, we are still waiting for our leaders to produce the green goods they have been promising for decades. Isn't it about time they produced them?

Demand Green Taxes. That's right: don't hide from green taxes, demand them! Taxing companies and individuals for contributing towards climate change is not the answer to climate change itself, and we certainly cannot rely on them for any long-term solution, but it is certainly an excellent step to take in the first stages. If we can deter businesses from excessive pollution, and individuals from wasteful lifestyles, then not only will it begin to reduce the uses of fossil fuels but it will also educate more people. People may be encouraged to use public transport with congestion charges in cities (as pioneered in London) and higher petrol duties, as well as perhaps subsidies for public transport. Drivers may be encouraged to buy more fuel-efficient cars by 'taxing out' of the market wasteful 4x4's and other gas-guzzling vehicles.

The taxes have to be daring: most heavily polluting businesses are taxed already, but the cost of the tax is so low that it is a burden they can live with, and does not put them off. It can be cheaper for some businesses to keep polluting and paying the little amount of tax than it is for them to change their operation and reduce the pollution. We cannot hope to avert widespread climate change in the next half century, and beyond, without a deep and concerted change to the way we exist on this planet. Therefore, we cannot rely on green taxes just as we cannot simply rely on converting all our fossil fuel power plants to renewable ones – every piece of the puzzle is needed to reach the lasting solution.

End Logging. Go one step further and plant trees instead, in an effort to combat your own carbon emissions. There is only one perfect and flawless way of taking carbon dioxide out of the atmosphere and it is embodied in something we see everyday in the world around us. Trees are wonderful, not only because they suck CO_2 in but because they give us oxygen, improving our health and respiration. Several organisations exist that plant trees for you for a small fee. Though afforestation is essentially the lazy man's way of combating climate change, and certainly not the whole answer, it certainly helps.

You can neutralise, with one tree, the CO_2 you produce in a year with heavy use of a computer, hi-fi or a television. Two trees can neutralise the average Rich Northern lifestyle for one month. 12,000 miles in you car can be paid for with the planting of five trees and you can run your home for a year, guilt free, with around eight trees. Try to pick a forestry organisation that plants only indigenous species of tree because this also helps the local ecosystems regain their former glory and return forests to their ancient wild state. Every tree one helps to reduce CO_2, improve our visual environment, reduce the spread of noise and dust, create habitats and reduce erosion of topsoil.

See:

Forest Ethics *– http://www.forestethics.org, for information on using market power to persuade and educate consumers, protect forests, reform forestry practices and increase power of indigenous communities.*

Global Forest Watch *– http://www.globalforestwatch.org to track corportations and government agencies.*

Pester politicians. Lobbying (or pestering) is easy. Local politicians can be contacted through letters, emails and in person at local surgeries. Ask them to make changes that will help prevent climate change, such as higher road tax on the least fuel-efficient cars, investment in large-scale renewable energies, and protection of forests and natural areas. Climate change is a profoundly political issue.

It can be disheartening trying to appeal to your government representative, and the process can be long. Often politicians can be quite dismissive about issues, but one thing is for sure - politicians only move to the beat of many drums, not just one. Get other people lobbying them too; get a group organised or tell your friends to get active. There is nothing more terrifying to a politician than seeing a swarm of their own constituents demanding change openly and publicly, and gaining support like a rolling snowball.

Convert military money to better uses. Every year world leaders spend around $825 thousand million on the military; the US alone spends around $320 thousand million of that total, more than the next ten biggest military spenders put together. Some countries spend so lavishly on their military that a fair proportion of their GDP is consumed in this manner – in 2001, Angola spent 17 per cent of its GDP on the military, Eritrea spent 21 per cent, Oman 14 per cent and Saudi Arabia 14 per cent. US military spending increased by almost a third between 1998 and 2003.

For some nations, a strong military keeps the country as an important player on the world stage, whether they really are or not; for other countries the military is just a large-scale job-creation scheme. Literally trillions of dollars are spent every single decade by states on their military. Nuclear weapons programs alone cost astronomical amounts of money. The estimated cost of the UK replacing and maintaining its Trident nuclear program for thirty years is around £76 billion. Analysts think that this money could go a long way to putting the country on the right path to cutting its carbon emissions. Oxford University's Environmental Change Institute predicts that every single UK home could be overhauled and made virtually carbon-neutral with just £5bn.

When we talk about getting the world economy off the fossil fuel drug we have to consider the costs of changing established infrastructures, electricity production, transportation, industry, new technology, changing peoples habits and so on. It all costs a lot of money. It would be naïve to think that we could simply do away with our militaries and the arms trade but is it really necessary that we spend so much money on it? Do impoverished African countries really need to spend more than a fifth of their GDP on weapons and soldiers? Does the US really need to spend more on its military activities than the next ten highest spenders put together? Do nuclear weapons prevent terrorism?

With all the poverty and environmental degradation out there in the world, the fact that the human race is still full of mistrust and infighting just reflects how little we've come from the bloodthirsty, civilisations of the ancient past. In fact, the only thing that has really changed is the technology.

Take to the streets. To a certain extent, marches and demonstrations can help to bring about change. There is nothing more symbolic for any major movement than millions of people taking to the streets with placards and songs and marching for change. It is a great visual representation of a movement and it can raise awareness amongst people who prefer not to do things unless it's very popular.

Direct action. Demonstrations can only do so much, and in nearly all cases it does virtually nothing. There was a massive anti-war movement around the world prior to the 2003 invasion of Iraq, involving tens of millions of people in several countries, but the war went ahead anyway. Part of the blame lay with the media, which failed to report accurately and fairly the extent of the anti-war movement. Politicians just ignored it.

Direct action *prevents* the media and politicians from ignoring it. Mohandas Gandhi didn't manage to drive the British from India by walking around in a large crowd and singing songs - he took peaceful direct action. Martin Luther King Jr. organised people to systematically break the unjust laws of the country and go to jail for it. The economy has to be disrupted to make any real mark on people's consciousness, whether it's organising a nationwide strike of teachers, lorry drivers, etc. or just taking a large group of friends and sitting down in the middle of a busy city street. Rapid change can only be found with practical displays of disquiet like this.

---

[i] Smith D. (2003). *The State of the World Atlas*. Earthscan Publications Ltd, London. Pg 129.

[ii] http://www.futureforests.com, Accessed 28th May 2006.

[iii] Ibid.

PART 5 - the Politics of Climate Change and the Future of Planet Earth

Chapter 14 – The Frog in the Water

There is an old saying that comes to mind whenever I see climate change being discussed by a climate sceptic, and the more time passes, the more the saying becomes relevant.

If you throw a frog into boiling water it will jump right out. If you put the frog into cold water, and gently heat the water up over time, then it won't sense the temperature change and just sit there, slowly boiling to death.

Although this is a gruesome image, and I'm not at all sure if it is true (I've never had the motivation to go out and see if I can con a frog into being boiled alive), the message is quite clear. If places and circumstances change very quickly then people are often driven to adapt, or oppose those changes. The rise of Marxism amongst large numbers of people throughout Europe, and the victory of the Bolsheviks during the Russian Revolution, was a force that gathered en masse over only a couple of decades, and was quickly matched by the rise of fascism – it's political opposite – a couple decades later in Hitler's Germany, Mussolini's Italy, and Franco's Spain. However, if something happens bit by bit, slowly and gradually, people are desensitised to it.

It's like being in a boat; you may only be floating at a few metres per minute away from the shore, but if you close your eyes and avert your attention elsewhere for long enough, then before you know it you are stranded in the middle of nowhere with no sign of land. That is what is going on right now. We saw climate change coming decades ago, but all this while we have done virtually nothing to challenge it. We've averted our eyes, become distracted, looked at the problem and delayed our response. People left it for their children to deal with, who left it for their children to deal with, who now sit here facing a family legacy none of them want to have.

If we had suddenly gone from the world of 1950 to the world of today then we would have noticed the changes straight away and immediately got into a panic. But we haven't, and instead we've sat around dealing with other things, satisfied in the knowledge that a few conferences here and there will sort it all out. We've spent almost 40 years (or even more) arguing about whether it is really going to happen, and then arguing about how we can stop it without spending too much money. Some are still sat around and can't feel the heat at all. And all the while we are still drifting from the shore, boiling alive in the pan.

Environmental issues are still political fodder

Politicians have their own interests to think about. It doesn't matter whether a politician cares about the environment; if the voting public aren't bothered then they won't be either. In Rich Northern countries this has always held back political solutions to environmental crises, particularly climate change. For decades the public were told that climate change was an issue for the far distant future, and so most people held no particular interest in keeping things like carbon emissions controlled - there are simply bigger issues to deal with: healthcare, police numbers, salaries, taxes etc. Politicians lending more than a half-hearted ear to the environmental problem found themselves wading into the shallow end of the voting pool.

In the 1980s and 1990s, as the whole issue progressed from scientific periodicals to the popular press, politicians began to realise that the implications of climate change could be potentially disastrous. Even British PM Margaret Thatcher got caught up in the emerging predictions about a bleak future. In 1982, at the outbreak of the Falklands War, Thatcher spoke with glee when she said: 'It's exciting to have a real crisis on your hands, when you have spent half your political life dealing with humdrum things like the environment'[i] – no words have better summed up politician's attitudes to environmental problems. By 1989, she had changed her tune considerably, saying: 'No issue will be more contentious than the need to control emissions of carbon dioxide ... Each country has to contribute and those countries who are industrialized must contribute more than those who are not.'

Unfortunately, just a few weeks later at a global warming conference in the Netherlands, the British government opposed Dutch proposals to stabilize and cut emissions. But the British were not the only government walking a different walk and talking a different talk.

Today, as climate is physically seen to change year after year, and evidence mounts in favour of human-induced climate change, more people are calling for tough action to tackle it. Politicians are being persuaded, not because they care in their hearts but because it can win votes. In the UK, the Conservative party transformed itself into a 'green' party, offering itself as the only party to take climate change seriously. But by its very definition a conservative political outlook cannot tackle climate change effectively. The solutions the party offered were virtually no improvement on what was already around, and in the background the old rhetoric of 'faith in capitalism' hung around hypocritically.

Around the world the same problem is occurring. Climate change is being taken into the party political arena, and though this may seem to be a good way of moving action along through a bit of healthy competition, it actually serves to bog real progress down with all the arguing and the irrelevant particulars of legislation. Governments seem reluctant to take full responsibility for the part they play in the issue. They set themselves poor targets that claim to be bold steps and turn the attention of journalists onto the consumers. Instead, the focus is put onto the average person and what they can do to reduce energy use in their homes. The real issue – energy production – remains outside of the spotlight.

Even in the most environmentally progressive countries, targets are flimsy. The UK government recognised, in a white paper of 2002, that it needed to create a low-carbon transport system to quash the high volume of CO_2 emissions that come from road transport. It envisaged that ten per cent of all new car sales will be emitting less than 100g/km of CO_2 at the tailpipe by 2012. At the time, a new car emitted around 178 grams of CO_2 per kilometre, and the European Voluntary Agreement aimed for this figure to be 140g/km by 2008, so the UK government target of 100g/km by 2012 was a reasonable improvement. But the target only calls for ten per cent of new cars to match this requirement, meaning that nine out of ten new cars sold in the UK can still produce more than this amount a whole decade after the white paper was published.

Such targets aim far too low. If the UK government was serious about tackling climate change it would pour more money into each little sector of its whole climate change plan, demand stricter legislation for the private sector, and set itself tougher targets on such things as energy production and transport. The same story is repeated again and again throughout the world. But it is not the nature of our current world economic system to seek large upheavals or revolutions; it likes slow and gradual change, things that the markets will not notice.

We can see the dangers of only solving the cheapest problems, and waiting for the more expensive ones to become cheaper to solve, in the example of air pollution. Smokestack sulphur dioxide clouds are still being produced on a huge scale, causing the problematic Global Dimming effect. And it isn't just a growing threat to the African monsoon belt. If the same dimming were to hinder the Asian monsoon, over 3,600 million people would be affected – leading to mass migrations on a scale never seen before. Clearly the dimming process has to be stopped, and thankfully many governments are introducing measures to tackle sulphate pollution already.

Governments are quite happy to tackle the sulphate problem because it is quite cheap to do so - like it was cheap to ban the use of CFCs in order to protect the ozone. However, evidence shows that the global dimming effect actually helps cool the global warming effect, masking its full potential. If governments were to make huge cuts in SO_2 pollution, *and not CO_2 at the same time*, then they would be removing this mask and revealing the true extent of global warming, thus unleashing an even worse case scenario than what is predicted at the moment.

If we abruptly stopped all our SO_2 emissions tomorrow, precipitation in the atmosphere would wash out the excess sulphate particles within a few weeks – hence the dimming effect would not linger. The warming effect would be boosted severely, depending on how effective the SO_2 dimming was at counteracting the warming beforehand.

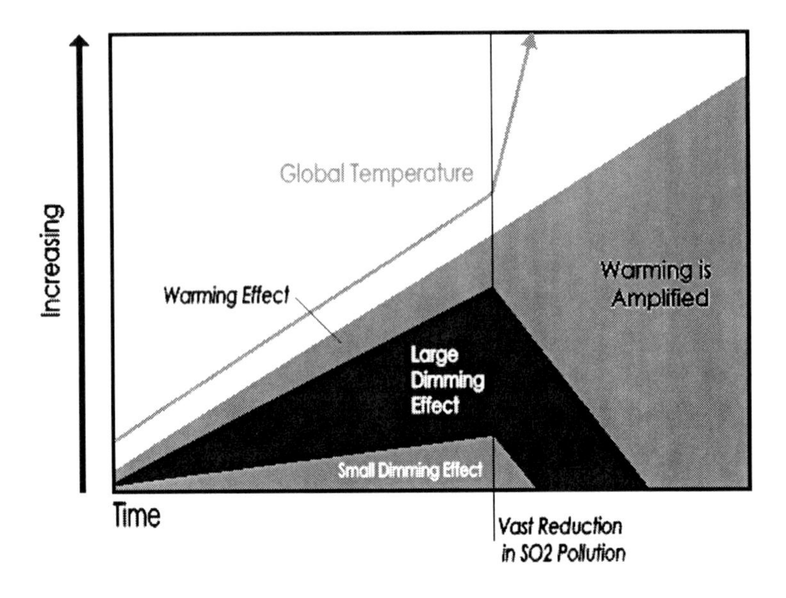

Figure 14.1 – If the Global Dimming effect of sulphur pollution is small then tackling it will only amplify global warming slightly. If the effect is larger, then global warming will be unleashed fully, sending global temperatures even higher.

If the dimming counteracted the warming substantially then a further dilemma may pop up. If, for example, we suddenly stopped all our carbon dioxide emissions abruptly, the gas would still linger in the atmosphere for hundreds or thousands of years. Consequently, even if we shut off all SO_2 *and* CO_2 emissions tomorrow, the warming would continue until the delay in the climate has caught up with previous CO_2 emissions (the delay is a couple of decades or thereabouts), before beginning to fall away slowly over the next few centuries, as CO_2 is naturally removed from the atmosphere. In effect, if the SO_2 pollution helps to counteract global warming more then the boost in warming would only be bigger. We can only hope that global dimming affects global warming as little as possible, and that the warming we are expecting to experience - thanks only to greenhouse gases - is as bad as its going to get.

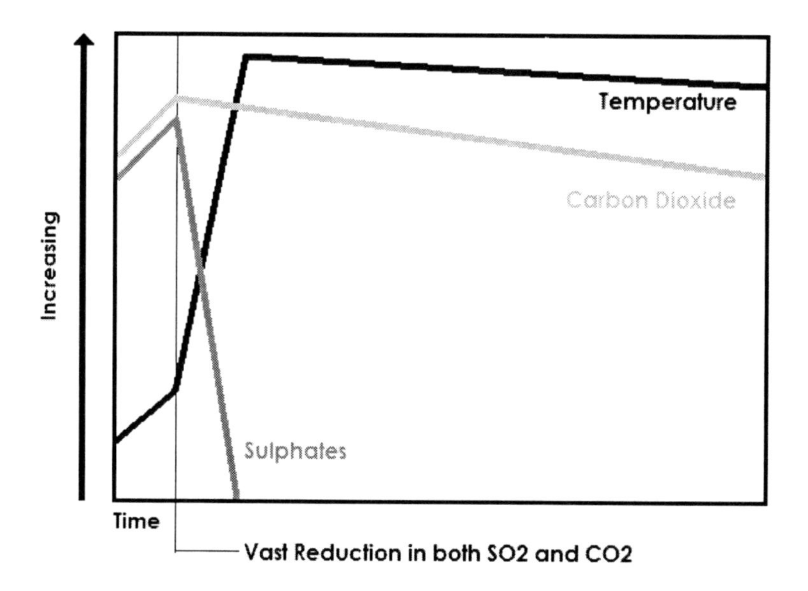

Figure 14.2 – If the dimming effect is large, and we eliminated both CO_2 and sulphate emissions at the same time, the warming of the CO_2 could linger for hundreds of years, while the cooling would not. This graph supposes that we eliminate both instantly.

The answer to this conundrum is tricky, since it seems that even instantly wiping out all of our CO_2 and sulphate pollution will still produce a horrible global warming effect. But this is better than stopping sulphate pollution and continuing to produce CO_2. It seems we have to tackle both at the same time, and hope that the Global Dimming effect is only small. Luck might be on our side. According to recent studies, the dimming effect could possibly be quite moderate. Then again, who knows what tomorrow's studies will tell us?

And hence, we cannot simply tackle the cheapest problems first and take our time over more expensive ones. As the dimming and warming relationship shows, being tight with our money isn't going to do us any favours.

Fossilised Politics: the seedy underbelly of climate change

Right about the time Margaret Thatcher was having her ecological epiphany (or probably before that) the big players in the fossil fuel industry were aiming to step up their game. Since the 1970s, when climate change made the mainstream media for the first time, anybody with even a remote stake in fossil fuels had been worried. For years the companies had been sending in their own scientists like spanners into the machine, trying to bring a halt to the idea of cutting carbon.

By 1990, they were leaning heavily on the IPCC as it tried to sort out its final draft of the first assessment report. In the end, they managed to get the wording watered down and the urgency and certainty behind the science severely weakened. They carried on throughout the 1990s, pressurising serious scientists and hounding politicians, under such harmless guises as the Global Climate Coalition, which - like some mafia front – was basically a flimsy curtain behind which lurked the most powerful names in the fossil fuel empire.

The climate sceptics didn't have to be very clever either, just very loud. Whereas serious scientists often found it hard to verbalise their findings or explain them in writing the media could understand, especially when conclusions obviously involve some degree of uncertainty, the sceptics had learned how to be noisy and get their message across in a language the people could follow. They got onto radio shows and television programmes and aroused the American public in a wave of suspicion at climatologists' findings. It's all very simple when you think about it. If you're a scientist, everything you say has to be backed up by lots of data and some sturdy evidence; if you're a critic all you have to do is say 'no' and reiterate that no one can predict the future.

The use of scientists by corporations, to put across a certain point of view, is a common public-relations strategy, known as 'the third party technique'. But, for all intents and purposes, these scientists may as well be actors, just voicing what they are being paid to voice. It's the craft of putting *your* words in someone else's mouth.

The sceptics then turned up the heat following the IPCC second assessment report of 1995 and began to wield their political power in bigger strokes. In the US, politics is quite different from the rest of the world. There are merely two political parties, and both live in the shadow of the other. Both take climate change like any other environmental problem – very lightly. American politicians were easily enticed by groups like the Global Climate Coalition or individual members, and many became more vocal against any form of climate change legislation, science or research.

Following the big let down of Kyoto (which was ratified by most countries only after a considerable delay, and produced targets that were far too poor to make any real dent in climate change) and the Byrd/Hagel saga surrounding the US position in Japan, climate change became something of a taboo issue on Capitol Hill. Daphne Wysham, a researcher at the Institute for Policy Studies, noted that the World Bank was flagrantly neglecting its stated remit of alleviating poverty and promoting sustainable development by spending 25 times more money on fossil fuel projects than it did on renewable energy projects. She also found that, for every nine out of ten energy projects that the World Bank funded, at least one corporation based in the seven richest countries benefited.[ii]

But, when people like Wysham tried to highlight such contradictions in the US Senate, they found that any information that happened to mention the words climate change would be stricken from the record. When you consider that the US is the largest stakeholder in the World Bank it's not hard to see what's going on.

Famously, in the 2000 presidential election, vice-president and Democratic candidate Al Gore was advised to focus less on climate change because it would turn voters off (though he and the party deny this). Gore lost the White House and took a step down from politics. Six years later – and whilst I was still writing this book – Al Gore went back to his passion and released a documentary about climate change, named 'An Inconvenient Truth', which was based on a series of lectures he had given the previous year in numerous places all around the world.

The 2000 presidential election may perhaps prove to be the biggest turning point in world history. The two candidates could not have been more different when it came to green issues. In one corner was Al Gore, the man who had studied under the famous Roger Reville (a main character in the story of the discovery of climate change), and who had already written a book about the environment called *Earth in the Balance: Ecology and the Human Spirit* in 1992.

In the other corner sat a staunch republican; a man who had owned or ran several oil companies before politics, who had countless friends and allies in fossil fuel tycoons, the oil-rich nation of Saudi Arabia and in other business leaders who had great interests in undoing environmental laws and carrying on 'business as usual'. If history had been different, and Gore had won the White House, the world today could have been a vastly different place. The great turning point of modern times went against the favour of the environmental movement, condemning the world to years of denial during the crucial period when attitudes towards climate change should have been changing the world over. As a result, future generations may look back at the 2000 election as the moment when our generation slipped up.

And we must also take into account the role that the American media has played in the last decade, transforming climate change from a serious debate about 'what to do to stop it' to a time-wasting 'are humans causing it at all.' In Europe this trend is beginning to wane, and more and more media outlets are getting behind climate change as a global campaign and a matter of urgency. However, in the United States the big media outlets are reluctant to see climate change as science-fact, and often ignore the issue altogether. Since the US is the world's biggest contributor to greenhouse gas emissions, having Americans take the issue seriously and act upon it is absolutely vital to any efforts to stop it; the American media is therefore playing a key role in preventing American citizens from acting with their virtual blackout.

In his documentary, Gore makes some vital points regarding the opinion of the media versus the general scientific consensus. He highlights a study published in Science magazine by Dr Naomi Oreskes of the University of California. In her study, Dr Oreskes selected a random sample of 928 peer-reviewed[1] articles, which represented around ten per cent of the total, dealing with climate change that had been published in scientific journals over the last ten years. Of the 928 articles, exactly *zero* disagreed with the general consensus: that human beings were responsible for causing the climate to change.[iii] So they could say with confidence that climate change sceptics, even the scientists, were in the very extreme minority.

[1] Peer-reviewed articles are reviewed by other scientists of that and similar fields, thus giving such studies added credibility compared with those that have just been written and not assessed.

Another study was simultaneously undertaken that looked at articles on climate change published in the four most influential newspapers in the US: the LA Times, the New York Times, the Wall Street Journal and the Washington Post. Of another random sample consisting of 636 articles (almost 18 per cent of the total) they found over half (53 per cent) gave equal weight to the general consensus and the sceptic view. Consequently, the American people remain some of the least convinced about global warming as anyone on the planet, despite some of the best science on the matter coming from American scientists.

Mother Nature versus the Almighty Buck

Despite increasing environmental awareness during the twentieth century, our species still treats ecological issues as low priority. Every year, the health of the planet decreases and future prospects look increasingly grim. Yet, every year we learn more about conservation, about climate change and about the natural world - so what is happening to this knowledge to stop it getting through? Something is standing in our way and preventing us from successfully looking after this planet.

Take the example of our natural forests. The world's forests are being wiped out at a rate of more than one per cent per year and more and more countries are being forced to clear their natural heritage for timber sales and agricultural land. But we already know that our forests are vital players in locking carbon away from the atmosphere, and their loss is our loss also.

For ten years during the 1990s, Western governments discussed protection of the Amazon rainforest, the world's biggest forest and the destination of around ten per cent of atmospheric CO_2. By 2000, they had struck a deal. Did they offer billions of dollars in aid to alleviate poverty in Brazil, thus creating jobs and wealth and undermining the need for a growing population, and for that population to cut down the rainforest for farmland? No, they didn't. Instead they got Brazil to agree to preserve just ten per cent of its forest in exchange for funding from the World Bank's Global Environment Facility. Just ten per cent - with funding that could be cut off at any time.

Consider also how hesitant the politicians have been to make economic sacrifices to prevent global climate change, time and time again. At Kyoto, for example, world leaders could only agree on a 5.2 per cent cut on 1990 greenhouse gas levels for most of the Rich North – the primary perpetrators of climate change. There was such a lack of self-sacrifice that the resultant targets drawn up were basically a large invitation for the world to keep on warming. In fact, the most enthusiastic endorsements of tough cuts only came from the Alliance of Small Island States, who faced losing their entire nation under the sea. Samoan Ambassador H E Tuiloma Neroni Slade clearly saw the obstacle standing in the way of bigger, more meaningful cuts when he said: 'The strongest human instinct is not greed - it is not sloth, it is not complacency – it is survival... We will not allow some to barter our homelands, our people and our culture for short term economic interest.'[iv] His words apparently didn't tug hard enough at the heart-strings of some delegates, who went on to suggest that it would cost a lot less to evacuate these islands than to cut emissions.

Kyoto targets are so poor that they actually give us a stark reminder of how dominant the big bully of economy is over the natural world. For example, Australia is one of the worst greenhouse gas emitters on the planet, generating 90 per cent of its electricity by burning coal.[v] However, Australia doesn't need to burn coal because, not only is it home to 28 per cent for the world's uranium and the best geothermal site anywhere, it is also blessed with plenty of bright sunshine and, in some parts, gusty and consistent winds. Australia only chooses to stick with coal because it is extremely cheap for them, and harnessing any of the other resources on a large scale would be slightly costly; money over matter.

In October 2006, a special review on future greenhouse warming was published by former World Bank chief economist Sir Nicholas Stern, outlining the economic fallout of more intense and more frequent weather catastrophes in a future dominated by climate change. The report rocked the European Union to the core, largely because it delved into the *third* front on the war against climate change that few people had actually considered to exist, let alone had bothered thinking very hard about. For decades we had pondered the effects of a rapidly changing global climate on human populations, and tried to speculate how the natural world would fare when faced with endless threats of extinction. But the actual *economics* behind future climate change had always been a grey area.

Consider this: Asian money markets currently hold about 60 per cent of global trade. Japan and the Eastern coastline of China and Korea are on a dangerous weather front, with threats from flooding, drought and heftier cyclones. Will a prolonged period of nasty weather events in East Asia threaten, if not devastate, global market economies?

In fact, the idea that tackling climate change is going to be incredibly damaging to the economy is entirely false. Many financial analysts now argue that becoming more energy efficient, turning on alternative energy supplies and tackling basic infrastructure problems will benefit economies on a whole, both in terms of extra employment and savings through self-sufficiency. It cannot simply be said that doing something about climate change is going to be expensive without having something else to compare it to – the cost of doing *nothing*. One of the best ways to calculate this future cost is to look at the past, and thanks to the insurance industry we can do that accurately.

The world's insurance industry has been growing more and more anxious over the last forty years due to increasing natural disasters losing them money. The El Nino year of 1998 alone cost insurers $89 thousand million on weather-related occurrences – that's more than the entire 1980s put together.[vi] Since the 1970s, insurance losses have risen by ten per cent every year, meaning that the total bill doubles every decade. Such an exponential increase in losses has the potential to devastate our world economy by as early as 2025 and could leave it in ruins by 2060.

What do the companies that insure the insurers make of all this? Munich Re is one of the world's biggest re-insurers and they came up with a rather conservative estimate for the annual damage bill of climate related events. Instead of the ten per cent increase every year, they suggested that the bill could reach a mere $500 thousand million in 2050, but admitted that even this is an increase they doubt the industry can absorb.

The Stern Review likens the need for the world to spend money on stabilizing emissions to an investment, a savvy one at that:

In broad brush terms, spending somewhere in the region of 1% gross world product on average for ever could prevent the world losing the equivalent of 10% gross world product for ever... (Stern, 2006).[vii]

So how much would it cost just to sit back and take climatic disasters on the chin? According to the Review, a 2°C increase in mean world temperature (a number considered a likely minimum) would cost 0.6 per cent of gross world product, though the upper end of the guess could be as high as four per cent. In today's terms, 0.6 per cent is around $390 thousand million. For a three-degree rise, the cost leaps to 1.4 per cent, with the worst case being 9.1 per cent. Once the temperature rises by 5°C, the mean expected cost is around 4.5 per cent, though it could cost as high as 23.3 per cent – that's virtually a quarter of all wealth produced on the planet... just to pay for the damage we've caused creating it.[viii]

It may seem at first that production that costs just a quarter of its final profits will still return a healthy three-quarter margin. In other words, why grumble about spending twenty-five cents producing something that gives you a dollar? But spending a quarter of all generated wealth on tidying up after ourselves is not only bad economic sense, but it will cost hundreds of millions of lives through lost food production and services, especially under a 5°C warming. Furthermore, when you factor in that it will cost less to stabilise all greenhouse gas emissions than to foot this large clean up bill (with the suffering and death attached) the argument is terminated.

Stern caused a stir. The issue suddenly seemed more urgent and more real; not to the ordinary working person, but to the industrial tycoons and business CEOs who had never previously considered global warming to be a threat to their bottom line. In response, the British government and European parliament jumped into the debate and promised tough new changes to help curb emissions of greenhouse gases, particularly carbon dioxide. The race was on to find a more effective solution than had already been set in stone at international conferences such as Kyoto. In Britain, the critics leapt at the idea, pointing out that UK emissions account for less than two per cent of the world's total, and that if the country stopped emitting altogether the benefits to the climate would be wiped out in less than two years by China's emissions alone. They also argued, quite rightly so, that the UK cannot tackle global climate change alone in a world where everyone from China to Chile is seeing their greenhouse gas emissions rise.

Still, the UK government announced its revolutionary strategy that it promised would minimise carbon emissions in the country and inspire other nations around the world to follow suit, therefore bringing an end to the climate problem once and for all, sometime in the near future. Prime Minister Tony Blair himself said that Britain wanted to lead the world into this new revolution. So what were the changes that were going to spark this new low-carbon society? Small taxes! The Environment Secretary announced that he wanted a £5 increase in air passenger duty to raise £400 million per year; petrol duty increases even if fuel prices fall, and a "substantial increase" in road tax for 4x4s.

European Commission president Jose Manuel Barroso also warned of the urgency to tackle climate change as quickly as possible, and proposed tough new laws on energy efficiency to cut energy costs by £70,000 million a year, and affect everything from water heaters and computers to removing the pointless standby button from all television sets. With sweeping taxes to be introduced as the first step towards a greener world, it seemed the average consumer was going to foot the bill for the greater climate problem.

It is an age-old ploy of the economy to lump added costs onto the consumer. We see it today in fair trade goods. Rather than pay the poor producer more money and settle with smaller profits, the big companies put the costs onto the consumer by increasing the selling price. In other words, the consumer pays more so that the world can be a fairer place. The answer to climate change is now coming out the same; if you want a greener world, you are going to have to pay for it. The industries will keep on making their profits, with all extra costs incurred onto the buyer. The market economy refuses to pay any of these costs. It all comes down to this: if you want a clean, green, safe future, without poverty, disease, pollution, malnutrition, and ill health, then you have to pay *more* for it. It is actually cheaper to live *with* these problems than without them! Why?

When did such things as a safe, green and peaceful future stop being the underlying purpose of humanity and start being a luxury that can be bought and sold? Additionally, why are we so anaesthetized in the face of the global market? Why are we slaves to this economic phenomenon that we ourselves created? How did we let capitalism become the Frankenstein monster that we stopped controlling the moment we created it – determining whether or not we can save the world?

Waiting for capitalism to slowly introduce an answer to climate change, and any other environmental catastrophe, is going to take too long. In China, there are plans to build hundreds of coal-fired power stations over the next few decades, simply because the cost of coal is so cheap. The story is much the same in many poor and developing countries. The system has determined that we will indeed keep on pumping out CO_2 emissions and we will do so as much as we can until the price of fossil fuels soars beyond economic sense, way off in the future when it's already too late. We're not going to make the investment hinted at by the Stern Review because all we really care about are our bills, not our legacy.

Economists and politicians are struggling with the idea that renewable energy is profitable and makes good financial sense. However, while the costs of fossil fuels are so cheap there is little likelihood that renewable energy is going to experience the much-needed boom that environmentalists are hoping for. Capitalism, by its very definition, cannot allow a resource to be discarded solely on its environmental damage, especially one that is worth so much money to those that control it. We cannot wait for fossil fuels to be priced out of the market – by then it will be too late. The very mechanisms of the world's financial system are wrestling with the mechanisms of the climate system and, though both waver, in the end only one of them is permanent.

Over the next two decades we may see the price of oil push well beyond its previous record high. The extraction of oil is more like squeezing treacle out of a rock than lifting bucketfuls of water from a river. When oil is first found there may be enough pressure to produce a 'gusher', where a fountain of thick black liquid is thrown into the sky. Only 15-25 per cent of oil can be recovered under its own pressure like this. For 30-45 per cent recovery, you have to pump water and natural gas down into the rock to maintain the pressure artificially. Finally, for up to 50 per cent oil recovery you have to inject heat or chemicals to make the oil less viscous. All this is very expensive, and can only allow about half of the oil reserve to be removed anyway, making the future prospects of oil recovery very bleak indeed.

The trillion-dollar fossil fuel industries, and others with financial fingers in many a fossil fuel pie, should sit up and take note. The best thing they can do right now is change to non-CO_2 sources of electricity; considering it an investment for the future. It will cost a fair sum to get the economy off its addiction to the black stuff, but the price of dealing with climate change will be many times higher for the global economy. Not enough investment, or none at all, and markets will begin to disintegrate. The average person will not be able to afford expensive cars, sending prices plummeting, and would tend to spend more on necessities rather than luxuries. Following energy markets, other markets would tumble too, including agriculture, fishing, services, financial institutions, manufacturing and so on. A small investment now could avoid all this.

The Fight Back

Supporters of the Kyoto Protocol's poor targets argue that even having small targets in place is better than having nothing. They also argue that criticising the Treaty is not progressive because it is the only treaty in place to take on carbon emissions. What would you replace Kyoto with, they ask. Well, commonsense dictates that small targets may well be better than nothing at all, but big targets are much better than small targets.

The problem is how to get governments to agree to tougher targets when they are already shuffling their feet to meet Kyoto? If world leaders can't agree to real targets for cutting emissions then it may come down to the power of the people having to force their hand. All around the world, hundreds of millions of people are sitting around watching their politicians squabble. Together, they amount to one big global force for change, stronger than anything we've covered so far. Even in countries where people are repressed from political activism, such as China, economic pressure from other countries will force repressive states to follow suit.

In the end, it's the action of the people that will decide the fate of the planet. Around the world, large groups of people have been massing to protest against the little political action actually taken by politicians despite plenty of words and promises. As a result, climate change has become a large mainstream issue with big and small pressure groups all over the world forcing it onto the political front, and demanding that their politicians take some action.

There has been a greater acceptance of climate change as a global threat even amongst the wider public, who are beginning to get behind the challenge. Yet politicians are quick to avoid action on curbing greenhouse gas emissions and changing fundamental keystones of the economy such as power plants and transport, particularly if they feel it may cost votes at the polls. But they should fear not, as it seems the public are already two steps ahead and are willing to make the sacrifices that the politicians think they won't.

According to a Guardian/ICM poll of 2006 (conducted in the UK with over a thousand individuals) 63 per cent of people were willing to pay higher taxes to tackle environmental problems, including climate change.[ix] Such taxes included energy wastage in the home, poor fuel-efficient cars and taxes on aviation, which cause a large proportion of CO_2 emissions at an increasingly faster rate every year. This followed an announcement in February 2006 by Prime Minister Blair that there would be no tax on air passengers, as he preferred to wait for better technology to come along and solve the problem. Though the government reiterated that its priority was a growing economy, rather than measures to tackle climate change, the same poll also found the public in disagreement; tackling climate change was third on people's lists of priorities, after improving the health service and improving education. Following that was security against terrorism and then a growing economy. Compare this poll with Blair's own words just a few months before: "The blunt truth about the politics of climate change is that no country will want to sacrifice its economy in order to meet this challenge."

Around the world, people are beginning to get edgy about global warming. In November 2006, more than 30,000 people took to the streets of London for the National Climate March, to oppose the government's lacklustre attitude towards the issue. Similar demonstrations took place all over, from Bermuda to Boston, Melbourne to Montreal, St Petersburg to Seoul and Bangladesh to Berlin. There was even a protest by Masai farmers in Nairobi. In Vermont, five hundred people got involved in about thirty actions in every corner of the state.

The Live Earth concerts of July 2007 kept this ball rolling. The seven-continent, 24-hour music extravaganza became the world's largest entertainment event ever – engaging roughly 2,000 million people. Greater than Live Aid of the 1980s, Live Earth sealed climate change as a mainstream issue, and simultaneously highlighted the vast numbers of people from around the world who were anxious about the climate crisis.

In July 2004, the American states of California, Connecticut, Iowa, New Jersey, New York (and New York City), Rhode Island, Vermont and Wisconsin filed suit against the five largest global warming producers in the US. These companies owned or operated at the time 174 fossil fuel power plants in 20 different states, emitting more than 650 million tonnes of CO_2

every year – about ten per cent of the nation's total. The legal action was made in an effort to reduce these companies' carbon dioxide emissions.

California Attorney General, Bill Lockyer said: "This lawsuit opens a new legal frontier in the fight against global warming – a challenge that poses a serious threat to our environment, our natural resources, our public health and safety, and our economy. A head-in-the-sand response is not an option. For the sake of our people and their future, we must act now. And requiring these major polluters to do their part is crucial to fighting the threat successfully."[x]

The fact that this suit was filed shows that even individual states and cities in the US are getting behind the struggle against global warming, despite what federal government says. It also highlights another significant turnaround in American environmental politics, whereby around 225 different cities in the country have implementing strategies of their own to reduce CO_2 emissions, below the targets called for in the Kyoto plan that the Bush Administration rejected. Such cities include: Los Angeles, San Francisco, Miami, Tallahassee, Atlanta, Chicago, New Orleans, Boston, Minneapolis, Las Vegas, New York City, Philadelphia, Vancouver, Salt Lake City, Honolulu and Washington D.C. itself.

"Our lawsuit is a huge, historic first step toward holding companies accountable for these pernicious pollutants that threaten our health, economy, environment and quality of life now and increasingly in the future," said Connecticut Attorney General, Richard Blumenthal, involved in the lawsuit. "The eventual effects of CO_2 pollution will be severe and significant - increasing asthma and heat-related illnesses, eroding shorelines, floods, and other natural disasters, loss of forests and other precious resources. We must act, wisely and quickly, to stem global warming - and safeguard both our environment and economy. Time is not on our side."

---

[i] Godrej, D. (2001). *The No-Nonsense Guide to Climate Change*. New Internationalist Publications Ltd. Oxford. Pg 88.

[ii] Wysham, D. (1999). 'The World Bank: Funding Climate Chaos,' *The Ecologist*, March/April.

[iii] Gore, A. (2006). *An Inconvenient Truth*, Bloomsbury Publishing, London. Pg 262-263.

[iv] Godrej, D. (2001). *The No-Nonsense Guide to Climate Change*. New Internationalist Publications Ltd. Oxford. Pg 105.

[v] Flannery, T. (2006). *The Weather Makers*. Penguin Books Ltd, London. Pg 226

[vi] Meyer, A. (2000). *Contraction & Convergence: the Global Solution to Climate Change*. Schumacher Briefing No.5. Green Books for the Schumacher Society, Devon. In: Flannery, T. (2006). *The Weather Makers*. Penguin Books Ltd, London. Pg 235.

[vii] http://www.hm-treasury.gov.uk/independent_reviews/stern_review_ economics_climate_change/stern_review_report.cfm, Accessed 25th Mar 2006, Pg 285. A physical copy can be obtained from Cambridge University press at http://www.cambridge.org/9780521700801.

[viii] Hope, C. (2003). 'The Marginal Impacts of CO2, CH4 and SF6 Emissions,' Judge Institute of Management Research Paper No. 2003/10. Cambridge, UK. In Stern Review: http://www.hm-treasury.gov.uk/independent_reviews/stern_review_ economics_climate_change/stern_review_report.cfm, Accessed 25th Mar 2006, Pg 295.

[ix] Adam, D. & Wintour, P. 'Most Britons willing to pay green taxes to save the environment,' *The Guardian*, Wednesday Feb. 22nd 2006.

[x] 'Eight States and NYC Sue Top Five US Global Warming Polluters', July 21st 2004. From: http//www.nyc.gov/html/law/downloads/pdf/pr072104.pdf, Accessed 23rd Jan 2007

Conclusion: What is the Ultimate Answer to the Climate Change Crisis?

The Holocene has so far proven to be the greatest gift our species has received from the planet. The combined stability and tranquillity of the last 10,000 years has provided humankind with a window of opportunity to go from small clusters of hunter-gatherers to huge swarms of insurance salespeople, public relations experts, football fans etc. We owe our entire recorded history – in fact, the very ability to record our history – to the Holocene.

The Holocene is incredibly unique, because not only is it stable but it is kind to our species. There are polar ice sheets but they are not so big that they suffocate our ability to occupy territory as high up as Norway and Alaska. The seas are low, giving us more land to live on and more coastlines. Because of where temperatures have come to rest, the human being can occupy the vast majority of the planet's land surface.

We are lucky that this period came into being, and considering the fragility of the global climate we are lucky that this period has not been ended abruptly by some climatic bump. The problem of human-induced climate change is therefore greater than just a problem for the next few generations. By twiddling with the climate system ourselves, we are in real danger of disrupting this fragile balance and sending the climate into a tailspin from which it will never find such stability again (at least not for hundreds of thousands of years). The climate has often been compared with a sleeping giant, which the human race is constantly poking at with sticks. Perhaps a better analogy would be a sleeping giant that we are constantly beating around the head with a bat, harder and harder as the years pass.

To end the Holocene because we couldn't manage our own success would be a disaster. As the saying goes, with power comes great responsibility. We now have the power to change the world. The question isn't 'how should we use this power?' but rather 'do we need this power in the first place?'

A possible path to take to buy us time might lie with the Poor South, and those carbon credits that the Rich North need so badly. The first step would be to enforce a worldwide, not-optional, clamp down on greenhouse gas emissions, with strict targets for every country to aim for. This is reminiscent of Kyoto, but with more urgency and a lot more strictness. An overall emission target should be set for the whole planet – for example, to emit no more than 100 billion tonnes of CO_2 between now and 2100. The next step would require that this total is split 50-50 between the Rich North and the Poor South – the reason why will become apparent in a moment.

Emissions targets should be unique to the individual nation, and reflect their current emissions divided by their adult population. Every country would then get a fixed rate of emissions and a specified length of time in which to use it up; so one country may have more time than others, depending on how bad they are now and how wasteful each person in society is. This gives the Poor South ample emission rights that they don't need. But instead of using them up over the rest of the century and trying to keep going business as usual, Poor South countries will sell these excess rights to Rich Northern countries who will probably need them desperately.

Yet this cannot work like the current carbon system, where emission rights are sold for money. Instead, the seller swaps the rights in exchange for alternative energy systems, reforestation programs, expertise and education programs, agricultural know-how, and other clean technologies – all paid for by the buying nation. This buys the Rich North the time it needs to convert into carbon free economies (not allow them longer to keep doing what they've always done) and allow the Poor South nations to 'leap-frog' the fossil fuel stage of development and go straight to the renewable energy stage.

Of course, all this would depend on making sure every country sticks to their targets and their trade agreements – unfortunately this is quite unlikely, especially when you consider most nations are going to fail at meeting their flimsy Kyoto targets. Enforcement of such a system would also be next to impossible, particularly when some nations have the power of veto at the United Nations and can therefore overrule, flout and ignore whatever rules they choose.

It shouldn't take another year like 2005 to get the ball rolling on tackling climate change properly, but it probably will do. If it does, we probably won't be ready for it any more than we were in 2005, we'll probably do nothing about it afterwards – again – and if we *do* it will probably be woefully inadequate.

The important thing to remember is this: we are doing *virtually* nothing about climate change. If a car salesman told you that there was around a 95 per cent chance that a car on his lot is going to break down the next day, you wouldn't buy that car. If someone told you there was a 95 per cent chance a train was going to derail, you wouldn't get on that train. If study after study showed that, unless everyone in your area started paying extra taxes, the area will be a cesspool of violence and crime for your children to grow up in, you would pay those taxes right away, or even demand new political change with different ideas. These are all commonsense outcomes for the scenarios; we all want to be safe, sheltered, looked after and to leave our children a better world than we ourselves had. Why then, with every climate study that is published stating that climate change is almost certainly going to strike in the next fifty years - and that we are the ones in control of it - do we carry on and do nothing. It lacks commonsense - it lacks passion and honour.

In 2007, a television documentary, shown on Channel 4 in the UK, attempted to kill off the idea of global warming. Brazenly entitled 'The Great Global Warming Swindle', the documentary explained any warming that the Earth had experienced over the last century or so was not caused by humans at all, but rather by the activity of the Sun. The argument focused around a couple of key 'facts'.

First, in 1991, the Danish atmospheric physicist Dr Eigil Friis-Christensen apparently found that recent temperature variations are in "strikingly good agreement" with the length of the cycle of sunspots. This was false, as a paper published in the journal Eos proved, showing that Friis-Christensen's data had been incorrectly handled. Friis-Christensen and another author then proposed that *cosmic radiation*, influenced by changes in the sun, was in strikingly good agreement with global cloud cover. Apparently this was warming the Earth. But, yet again, the data had been handled incorrectly, and in fact it didn't measure global cloud cover at all. A rebuttal paper in the Journal of Atmospheric and Solar-Terrestrial Physics shows that, when the right data are used, there is no correlation.

Friis-Christensen's co-author, Henrik Svensmark, declared that the correlation was with only *low* cloud cover, and that cosmic rays could be forming tiny particles in the atmosphere. The documentary virtually concluded that temperature rises were the result of cosmic rays.[1]

The second 'fact' was centred on a Professor John Christy and his assertion that the temperature measured at the Earth's surface didn't match the temperature of the lower atmosphere – and therefore the human emissions link wasn't actually there. But, what the documentary failed to mention was that in 2005 three papers in Science proved Christy wrong. The documentary also omitted the significant fact that Christy himself had admitted his mistake in a paper he co-authored in 2006: "this significant discrepancy [between surface and lower atmospheric temperatures] no longer exists because errors in the satellite and radiosonde data have been identified and corrected."

Journalists commentate on our world. They are the eyes and ears, as well as the mouths, for everything that is going on around us. More than this, they are the brain too – telling us things we didn't realise and bringing us to conclusions they might want us to see. If we rely on the media for honesty we forfeit our own ability of investigation and discovery. Environmentalism is a philosophical outlook. What could be more morally correct than to defend the natural world? The natural world provides us with our food, our shelter, our habitats, our climate, our air and our fresh water. Defending Gaia is the most fundamentally imperative occupation you could have. In fact, it is probably the most natural.

Essentially in our struggle, the market economy can no longer have the last word about our environment and our future. We cannot just tinker here and there with the market mechanisms; instead it needs a complete overhaul – a total revolution in the way the system works. Australia and the US are quite open about their reasons for not tackling climate change: the costs are too high. Instead they insist that a strong economy is the only way to ensure against future disaster. This is like seeing an asteroid hurtling to Earth, but instead of getting together on an international level and figuring out ways to divert it, you simply dig a hole in the ground, or build a wooden fort around your house, and hope for the best.

[1] For a point by point disproval of these 'facts' look at NASA scientist Dr Gavin Schmidt's jottings at: http://www.realclimate.org.

'No snowball in an avalanche ever feels responsible.' – Voltaire, 1694-1778, French philosopher.

The long-term solution can only lie with a massive sacrifice of world government money, and vast reduction on our demands on environment. It's hard to gauge from your home or office the exact scale of our species' impact on the planet, but it really is incredible. We simply cannot keep going the way we are going, and we cannot expect technology to answer all of our problems. In most cases, the *costs* of new technology that are given by the market add unwanted delays that prevent that technology from breaking out. This is why we're still burning fossil fuels, why alternative energy production is being neglected, and why governments invest more on trying to clean up fossil fuels than turning them off altogether. Projects like the Sahara solar pipe energy super-grid, which could potentially provide enough energy for most of Europe and the Near East, are shunned.

So what do we do? What do we sacrifice? If we want to turn this into another analogy then perhaps we could envisage a huge juggernaut heading full pelt towards the end of a cliff. By trying to carry on with the current economic system, and trying to prevent climate change through small changes, we are in effect pushing our foot all the way down on the acceleration and applying a tiny bit of pressure on the break. Instead, wouldn't it be better to take our foot off the acceleration, slam it down on the brake, turn off the engine, get out and throw the keys away?

Throughout this book I have purposefully avoided direct reference to the ethics of the how humans treat the natural world, particularly from a religious perspective. But to ignore the input of our ethics and our current religions would render the whole book pointless. Climate change is a problem for every society on the planet – how we deal with it comes down to our moral ideals – and every society on the planet has some sort of worldview, either religious or not.

First, from a religious point of view, there are several problems that pop up between humans and our complete harmony with nature. Religions - no matter how big or small - often deal with the concept of an after-life. It is estimated that around seventy five per cent of people on the planet believe in some higher-power[2], either multiple or singular. Christians, Muslims and Hindus alone account for more than half of all people on the planet. In effect, this is a planet of God-worshippers in some form (the word 'God' can envelope all manners of theism in this sense). Particularly in the largely related religions of Judaism, Christianity and Islam, the promise of an after-life is what brings peace of mind. When religious people die they expect eternal tranquillity, happiness and love. They imagine a place of beauty, of endless flowing rivers, lush green forests, crisp clouds or wide-open fields of green.

So what's wrong with that? As Karl Marx suggested, the promise of a wonderful after-life is the mechanism by which the masses are made satisfied with their lot in this world. By the same idea, we can see why it is so hard for our species to live in harmony with the natural world. Why preserve or improve the beauty of nature here on Earth if you're going to Paradise anyway? In his analysis of the phenomena of religion, 'The God Delusion', Professor Richard Dawkins put this point with a quote from Ronald Reagan's Secretary of the Interior: 'We don't have to protect the environment; the Second Coming is at hand.'

As far as the mainstream religions are concerned, the relationship between humans and the natural world is full of mixed messages. Are we stewards of this Earth, whose job it is to protect the environment, or is the environment there to be subdued, as the Old Testament suggests in Genesis?

Even if Genesis is only symbolic then what of the tale of the flood of Noah, whereby every living thing (apart from two of each species) was killed, including forests, mammals and other humans? God, it seems, created this amazing paradise for us, but then decided to completely destroy it all as a side-effect of revenge on disobedient humans. Perhaps *both* stories are symbolic, but even so, what kind of message is this symbolism trying to get across –that, so long as we are good people, nature is expendable?

Religion itself is not causing the greater environmental problem on Earth, but with its declarations that nature is here for *our* benefit and that a wonderful paradise will be waiting after we die, it is little wonder why many people refuse to take the demise of Gaia seriously - why many people refuse to stand up and defend Earth in its most needy hour.

[2] Of course, this discounts atheists, Buddhists, Confucians and other non-deity driven philosophies.

But of course, if we're going to prevent climate change from turning nasty on us then we need to change more than just those morals we receive from religion. It seems that human society has always drifted from one worldview to another, with increasing tolerance and understanding as times move on and we know more about the world. It was once quite commonplace for one group to obliterate nearby settlements for their own personal gain - today the act is (almost) unheard of, or at least frowned upon. Though George Washington owned his own slaves, Abraham Lincoln saw slavery as immoral, but stopped short of seeing non-whites as equal citizens with equal intelligence – something the American public would no longer accept in their modern presidents.

Times are always changing, and today's elderly people are often bewildered by what is deemed acceptable by modern standards. The Magna Carta was signed in 1215, the Emancipation Proclamation in 1863, and the Civil Rights Act in 1957. Once slavery was outlawed in most industrial nations, people started to demand the equal rights for women, with various movements popping up from the late nineteenth and early twentieth centuries. Native Americans were finally recognised as equals in the US with the Indian Citizenship Act of 1924. The Endangered Species Act of 1973 kicked off a whole new movement altogether. Of course, racial discrimination and slavery still exist today and women are still often seen as inferior to men. But to most people, racism and sexism are appalling acts, and we frown upon those states or individuals who commit them. Perhaps, one day before 2100, we will see the very last slave set free on this planet, or the complete worldwide acceptance of women as equals to men; one person equal to another.

Climate change fits into this changing moral world view. The current emerging movement is that of animal rights, which – though it has been around for decades now – is only just showing signs of being taken seriously. Intelligent and powerful individuals are starting to have deep discussions into how much pain animals feel, how intelligence is quantified in animals, and how rights might apply to some species, such as apes and domestic pets. Such a discussion would have been laughable *fifty* years ago, let alone five hundred. This build up suggests that one day most people on the planet will renounce violence or exploitation of animals. Perhaps nature as a whole is next.

What needs to happen to solve climate change is for our species to stop seeing nature as a complex collection of resources that we can use. Instead we need to realise that nature is as much a part of us as we are of it. We need to understand that exploitation and degradation of the environment affects our species too. While current generations have to live with environmental issues created in the past, such as inner city smog and melting glaciers, future generations will have to live with whatever we create now. Our fate does not exist outside of nature – we are bound to it and we can never break that bond. The suffering of the natural world at our hands will come back to our own suffering.

If a person is willing to regard Gaia as an equal to the human, then they will want to do all they can to prevent climate change from worsening for the sake of the planet and other life-forms. If they cannot regard nature is equal to humans then they should do all they can to prevent climate change from worsening for the sake of people. There is no third way. To not act on global environmental issues like climate change goes beyond selfishness, because to be selfish would be to *save yourself*; to not act at all is ludicrous.

"I am life which wills to live, in midst of life that wills to live." – Albert Schweitzer.

Our present global economic system is not centred on the environment. It's not even factored into the equation, other than as raw materials or resources for our use. Economists have been struggling for many years trying to decide how best to value the natural world into their equations. How much money is an African Elephant worth? Can we sacrifice one or two to use their land for agriculture? What about the climate? How much climate change will the market tolerate? At present, everybody knows such things are indeed valuable, but they have no actual financial value in essence. As a result, we lose elephants and change the climate. We catch too many fish and clear forests. The economic benefits – the human things we've already given financial value – nearly always win.

But can you put a price on the natural world? Can you put a price on a Humpback Whale or a Sloth? As soon as you do so, you remove their intrinsic right to life. Imagine doing the same for a human life ('Oh, its okay if fifty people lose their livelihoods and their homes… eventually dying – we can make a lot of money constructing this diamond mine!'). Most people just wouldn't consider it!

And again, what about the climate? Surely something so globally significant, so fine-tuned to the needs of every life-form on Earth, is priceless. Surely it is something we just cannot even consider harming.

This is the final and most important lesson in the whole climate change saga. Earth, our home, needs protection. It is magnificent. It is awe-inspiring. It is glorious. Even now, writing this, I'm struggling to describe it with words that seem adequate. Earth, Gaia, Mother Nature - whatever you want to call it – is the epitome of wonder. To ruin it for the sake of (an incredibly) short-term financial gain is deeply immoral and shameful. What message are we passing on to future generations? That we don't care for them? That we don't care for the planet?

One consolation I have when I think of the possible disaster ahead is that one day the Earth will return from its breakdown and flourish once more with life, forests and such. We may send it spiralling away into desolation for the next few hundred years, but in a few thousand it will bounce back and thrive once again. Life will find a way and the Earth will be around for a long time yet. What we have to decide is whether we will be there, too.

Dear Siona & David,

I hope you're well & wish you all the best luck in the world.

Love Always,

Saba

xXx

Printed in the United Kingdom by
Lightning Source UK Ltd., Milton Keynes
138261UK00001B/103/P